T0350146

Stochastic Approximation

This simple, compact toolkit for designing and analyzing stochastic approximation algorithms requires only basic literacy in probability and differential equations. Yet these algorithms have powerful applications – for example, in control and communications engineering, artificial intelligence and economic modelling.

The dynamical systems viewpoint treats an algorithm as a noisy discretization of a limiting differential equation and argues that, under reasonable hypotheses, it tracks the asymptotic behaviour of the differential equation with probability one. The limiting differential equation, which can usually be obtained by inspection, is easier to analyze.

Novel topics covered in the book include finite-time behaviour, multiple timescales and asynchronous implementation. A separate chapter gives a useful taxonomy of applications, with concrete examples from engineering and economics. Notably it covers several variants of stochastic gradient-based optimization schemes, fixed-point solvers, which are commonplace in learning algorithms for approximate dynamic programming, and some models of collective behaviour. Three appendices give self-contained summaries of background material from analysis, differential equations and probability.

Ideal for graduate students, researchers and practitioners in electrical engineering and computer science, especially those working in control, communications, signal processing and machine learning, it is also relevant to economics, probability and statistics.

Stochastic Approximation
A Dynamical Systems Viewpoint

Vivek S. Borkar

Tata Institute of Fundamental Research, Mumbai

CAMBRIDGE
UNIVERSITY PRESS

University Printing House, Cambridge CB2 8BS, United Kingdom

One Liberty Plaza, 20th Floor, New York, NY 10006, USA

477 Williamstown Road, Port Melbourne, VIC 3207, Australia

314-321, 3rd Floor, Plot 3, Splendor Forum, Jasola District Centre, New Delhi - 110025, India

79 Anson Road, #06-04/06, Singapore 079906

Cambridge University Press is part of the University of Cambridge.

It furthers the University's mission by disseminating knowledge in the pursuit of education, learning and research at the highest international levels of excellence.

www.cambridge.org
Information on this title: www.cambridge.org/9780521515924

Sold and distributed in all countries except India, Pakistan, Bangladesh and Sri Lanka by Cambridge University Press

Sold and distributed in India, Pakistan, Bangladesh and Sri Lanka by Hindustan Book Agency

First published 2008

A catalogue record for this publication is available from the British Library

ISBN 978-0-521-51592-4 Hardback

Contents

Preface

Stochastic approximation was introduced in a 1951 article in the *Annals of Mathematical Statistics* by Robbins and Monro. Originally conceived as a tool for statistical computation, an area in which it retains a place of pride, it has come to thrive in a totally different discipline, viz., that of electrical engineering. The entire area of 'adaptive signal processing' in communication engineering has been dominated by stochastic approximation algorithms and variants, as is evident from even a cursory look at any standard text on the subject. Then there are the more recent applications to adaptive resource allocation problems in communication networks. In control engineering too, stochastic approximation is the main paradigm for on-line algorithms for system identification and adaptive control.

This is not accidental. The key word in most of these applications is *adaptive*. Stochastic approximation has several intrinsic traits that make it an attractive framework for adaptive schemes. It is designed for uncertain (read 'stochastic') environments, where it allows one to track the 'average' or 'typical' behaviour of such an environment. It is incremental, i.e., it makes small changes in each step, which ensures a graceful behaviour of the algorithm. This is a highly desirable feature of any adaptive scheme. Furthermore, it usually has low computational and memory requirements per iterate, another desirable feature of adaptive systems. Finally, it conforms to our anthropomorphic notion of adaptation: It makes small adjustments so as to improve a certain performance criterion based on feedbacks received from the environment.

For these very reasons, there has been a resurgence of interest in this class of algorithms in several new areas of engineering. One of these, viz., communication networks, is already mentioned above. Yet another major application domain has been artificial intelligence, where stochastic approximation has provided the basis for many learning or 'parameter tuning' algorithms in soft computing. Notable among these are the algorithms for training neural networks and the algorithms for reinforcement learning, a popular learning paradigm for autonomous software agents with applications in e-commerce, robotics, etc.

Yet another fertile terrain for stochastic approximation has been in the area of economic theory, for reasons not entirely dissimilar to those mentioned above. On one hand, they provide a good model for collective phenomena, where *micromotives* (to borrow a phrase from Thomas Schelling) of individual agents aggregate to produce interesting *macrobehaviour*. The 'nonlinear urn' scheme analyzed by Arthur and others to model increasing returns in economics is a case in point. On the other hand, their incrementality and low per iterate computational and memory requirements make them an ideal model of a *boundedly rational* economic agent, a theme which has dominated their application to learning models in economics, notably to learning in evolutionary games.

This flurry of activity, while expanding the application domain of stochastic approximation, has also thrown up interesting new issues, some of them dictated by technological imperatives. Consequently, it has spawned interesting new theoretical developments as well. The time thus seemed right for a book pulling together old and new developments in the subject with an eye on the aforementioned applications. There are, indeed, several excellent texts already in existence, many of which will be referenced later in this book. But they tend to be *comprehensive* texts: excellent for the already initiated but rather intimidating for someone who wants to make quick inroads. Hence a need for a 'bite-sized' text. The present book is an attempt at one.

Having decided to write a book, there was still a methodological choice. Stochastic approximation theory has two somewhat distinct strands of research. One, popular with statisticians, uses the techniques of martingale theory and associated convergence theorems for analysis. The second, popular more with engineers, treats the algorithm as a noisy discretization of an ordinary differential equation (*o.d.e.*) and analyzes it as such. We have opted for the latter approach, because the kind of intuition that it offers is an added advantage in many of the engineering applications.

Of course, this is not the first book expounding this approach. There are several predecessors such as the excellent texts by Benveniste–Metivier–Priouret, Duflo, and Kushner–Yin referenced later in the book. These are, however, what we have called *comprehensive* texts above, with a wealth of information. This book is *not* comprehensive, but is more of a compact account of the highlights to enable an interested, mathematically literate reader to run through the basic ideas and issues in a relatively short time span. The other 'novelties' of the book would be a certain streamlining and fine-tuning of proofs using that eternal source of wisdom – *hindsight*. There are occasional new variations on proofs sometimes leading to improved results (e.g., in Chapter 6) or just shorter proofs, inclusion of some newer themes in theory and applications, and so on. Given the nature of the subject, a certain mathematical sophistication was unavoidable. For the benefit of those not quite geared for it, we have collected the more advanced mathematical requirements in a few appendices. These should serve as a source for quick reference and pointers to the literature, but not as a replacement for a firm grounding in the respective areas. Such grounding is a must for anyone wishing to contribute to the *theory* of stochastic approximation. Those interested more in *applying* the results to their respective specialties may not feel the need to go much further than this little book.

Let us conclude this long preface with the pleasant task of acknowledging all the help received in this venture. The author forayed into stochastic approximation around 1993–1994, departing significantly from his dominant activity till then, which was controlled Markov processes. This move was helped by a project on adaptive systems supported by a Homi Bhabha Fellowship. More than the material help, the morale boost was a great help and he is immensely grateful for it. His own subsequent research in this area has been supported by grants from the Department of Science and Technology, Government of India, and was conducted in the two 'Tata' Institutes: Indian Institute of Science at Bangalore and the Tata Institute of Fundamental Research in Mumbai. Dr. V. V. Phansalkar went though the early drafts of a large part of the book and with his fine eye for detail, caught many errors. Prof. Shalabh Bhatnagar, Dr. Arzad Alam Kherani and Dr. Huizen (Janey) Yu also read the drafts and pointed out corrections and improvements (Janey shares with Dr. Phansalkar the rare trait for having a great eye for detail and contributed a lot to the final clean-up). Dr. Sameer Jalnapurkar did a major overhaul of chapters 1–3 and a part of chapter 4, which in addition to fixing errors, greatly contributed to their readability. Ms. Diana Gillooly of Cambridge University Press did an extremely meticulous job of editorial corrections on the final manuscript. The author takes full blame for whatever errors that remain. His wife Shubhangi and son Aseem have been extremely supportive as always. This book is dedicated to them.

Vivek S. Borkar
Mumbai, February 2008

1
Introduction

Consider an initially empty urn to which balls, either red or black, are added one at a time. Let y_n denote the *number* of red balls at time n and $x_n \stackrel{\text{def}}{=} y_n/n$ the *fraction* of red balls at time n. We shall suppose that the conditional probability that the next, i.e., the $(n+1)$st ball is red given the past up to time n is a function of x_n alone. Specifically, suppose that it is given by $p(x_n)$ for a prescribed $p : [0,1] \to [0,1]$. It is easy to describe $\{x_n, n \geq 1\}$ recursively as follows. For $\{y_n\}$, we have the simple recursion

$$y_{n+1} = y_n + \xi_{n+1},$$

where

$$\begin{aligned} \xi_{n+1} &= \quad 1 \text{ if the } (n+1)\text{st ball is red,} \\ &= \quad 0 \text{ if the } (n+1)\text{st ball is black.} \end{aligned}$$

Some simple algebra then leads to the following recursion for $\{x_n\}$:

$$x_{n+1} = x_n + \frac{1}{n+1}(\xi_{n+1} - x_n),$$

with $x_0 = 0$. This can be rewritten as

$$x_{n+1} = x_n + \frac{1}{n+1}(p(x_n) - x_n) + \frac{1}{n+1}(\xi_{n+1} - p(x_n)).$$

Note that $M_n \stackrel{\text{def}}{=} \xi_n - p(x_{n-1}), n \geq 1$ (with $p(x_0) \stackrel{\text{def}}{=}$ the probability of the first ball being red) is a sequence of zero mean random variables satisfying $E[M_{n+1}|\xi_m, m \leq n] = 0$ for $n \geq 0$. This means that $\{M_n\}$ is a *martingale difference sequence* (see Appendix C), i.e., uncorrelated with the 'past', and thus can be thought of as 'noise'. The above equation then can be thought of as a noisy discretization (or *Euler scheme* in numerical analysis parlance) for the ordinary differential equation (o.d.e. for short)

$$\dot{x}(t) = p(x(t)) - x(t),$$

for $t \geq 0$, with nonuniform stepsizes $a(n) \overset{\text{def}}{=} 1/(n+1)$ and 'noise' $\{M_n\}$. (Compare with the standard Euler scheme $x_{n+1} = x_n + a(p(x_n) - x_n)$ for a small $a > 0$.) If we assume $p(\cdot)$ to be Lipschitz continuous, o.d.e. theory guarantees that this o.d.e. is well-posed, i.e., it has a unique solution for any initial condition $x(0)$ that in turn depends continuously on $x(0)$ (see Appendix B). Note also that the right-hand side of the o.d.e. is nonnegative at $x(t) = 0$ and nonpositive at $x(t) = 1$, implying that any trajectory starting in $[0,1]$ will remain in $[0,1]$ forever. As this is a scalar o.d.e., any bounded trajectory must converge. To see this, note that it cannot move in any particular direction ('right' or 'left') forever without converging, because it is bounded. At the same time, it cannot change direction from 'right' to 'left' or vice versa without passing through an equilibrium point: This would require that the right-hand side of the o.d.e. changes sign and hence by continuity must pass through a point where it vanishes, i.e., an equilibrium point. The trajectory must then converge to this equilibrium, a contradiction. (For that matter, the o.d.e. couldn't have been going both right and left at any given x because this direction is uniquely prescribed by the sign of $p(x) - x$.) Thus we have proved that $x(\cdot)$ must converge to an equilibrium. The set of equilibria of the o.d.e. is given by the points where the right-hand side vanishes, i.e., the set $H = \{x : p(x) = x\}$. This is precisely the set of fixed points of $p(\cdot)$. Once again, as the right-hand side is continuous, is ≤ 0 at 1, and is ≥ 0 at 0, it must pass through 0 by the mean value theorem and hence H is nonempty. (One could also invoke the Brouwer fixed point theorem (Appendix A) to say this, as $p : [0,1] \rightarrow [0,1]$ is a continuous map from a convex compact set to itself.)

Our interest, however, is in $\{x_n\}$. The theory we develop later in this book will tell us that the $\{x_n\}$ 'track' the o.d.e. with probability one in a certain sense to be made precise later, implying in particular that they converge a.s. to H. The key factors that ensure this are the fact that the stepsize $a(n)$ tends to zero as $n \rightarrow \infty$, and the fact that the series $\sum_n a(n) M_{n+1}$ converges a.s., a consequence of the martingale convergence theorem. The first observation means in particular that the 'pure' discretization error becomes asymptotically negligible. The second observation implies that the 'tail' of the above convergent series given by $\sum_{m=n}^{\infty} a(n) M_{n+1}$, which is the 'total noise added to the system from time n on', goes to zero a.s. This in turn ensures that the error due to noise is also asymptotically negligible. We note here that the fact $\sum_n a(n)^2 = \sum_n (1/(n+1)^2) < \infty$ plays a crucial role in facilitating the application of the martingale convergence theorem in the analysis of the urn scheme above. This is because it ensures the following sufficient condition for martingale convergence (see Appendix C):

$$\sum_n E[(a(n)M_{n+1})^2|\xi_m, m \le n] \le \sum_n a(n)^2 < \infty, \text{ a.s.}$$

One also needs the fact that $\sum_n a(n) = \infty$, because in view of our interpretation of $a(n)$ as a time step, this ensures that the discretization does cover the entire time axis. As we are interested in tracking the asymptotic behaviour of the o.d.e., this is clearly necessary.

Let's consider now the simple case when H is a finite set. Then one can say more, viz., that the $\{x_n\}$ converge a.s. to some point in H. The exact point to which they converge will be random, though we shall later narrow down the choice somewhat (e.g., the 'unstable' equilibria will be avoided with probability one under suitable conditions). For the time being, we shall stop with this conclusion and discuss the *raison d'etre* for looking at such *'nonlinear urns'* .

This simple set-up was proposed by W. Brian Arthur (1994) to model the phenomenon of increasing returns in economics. The reader will have heard of the 'law of diminishing returns' from classical economics, which can be described as follows. Any production enterprise such as a farm or a factory requires both fixed and variable resources. When one increases the amount of variable resources, each additional unit thereof will get a correspondingly smaller fraction of fixed resources to draw upon, and therefore the additional returns due to it will correspondingly diminish.

While quite accurate in describing the traditional agricultural or manufacturing sectors, this law seems to be contradicted in some other sectors, particularly in case of the modern 'information goods'. One finds that larger investments in a brand actually fetch larger returns because of standardization and compatibility of goods, brand loyalty of customers, and so on. This is the so-called 'increasing returns' phenomenon modelled by the urn above, where each new red ball is an additional unit of investment in a particular product. If the predominance of one colour tends to fetch more balls of the same, then after some initial randomness the process will get 'locked into' one colour which will dominate overwhelmingly. (This corresponds to $p(x) > x$ for $x \in (x_0, 1)$ for some $x_0 \in (0, 1)$, and $< x$ for $x \in (0, x_0)$. Then the stable equilibria are 0 and 1, with x_0 being an unstable equilibrium. Recall that in this set-up the equilibrium x is stable if $p'(x) < 1$, unstable if $p'(x) > 1$.) When we are modelling a pair of competing technologies or conventions, this means that one of them, not necessarily the better one, will come to dominate overwhelmingly. Arthur (1994) gives several interesting examples of this phenomenon. To mention a few, he describes how the VHS technology came to dominate over Sony Betamax for video recording, why the present arrangement of letters and symbols on typewriters and keyboards (QWERTY) could not be displaced by a superior arrangement called DVORAK, why 'clockwise' clocks eventually displaced 'counterclockwise' clocks, and so on.

Keeping economics aside, our interest here will be in the recursion for $\{x_n\}$ and its analysis sketched above using an o.d.e. The former constitutes a special (and a rather simple one at that) case of a much broader class of stochastic recursions called 'stochastic approximation' which form the main theme of this book. What's more, the analysis based on a limiting o.d.e. is an instance of the 'o.d.e. approach' to stochastic approximation which is our main focus here. Before spelling out further details of these, here's another example, this time from statistics.

Consider a repeated experiment which gives a string of input-output pairs (X_n, Y_n), $n \geq 1$, with $X_n \in \mathcal{R}^m, Y_n \in \mathcal{R}^k$ resp. We assume that $\{(X_n, Y_n)\}$ are i.i.d. Our objective will be to find the 'best fit' $Y_n = f_w(X_n) + \epsilon_n, n \geq 1$, from a given parametrized family of functions $\{f_w : \mathcal{R}^m \to \mathcal{R}^k : w \in \mathcal{R}^d\}$, ϵ_n being the 'error'. What constitutes the 'best fit', however, depends on the choice of our error criterion and we shall choose this to be the popular 'mean square error' given by $g(w) \overset{\text{def}}{=} \frac{1}{2}E[||\epsilon_n||^2] = \frac{1}{2}E[||Y_n - f_w(X_n)||^2]$. That is, we aim to find a w^* that minimizes this over all $w \in \mathcal{R}^d$. This is the standard problem of nonlinear regression. Typical parametrized families of functions are polynomials, splines, linear combinations of sines and cosines, or more recently, wavelets and neural networks. The catch here is that the above expectation cannot be evaluated because the underlying probability law is not known. Also, we do not suppose that the entire string $\{(X_n, Y_n)\}$ is available as in classical regression, but that it is being delivered one at a time in 'real time'. The aim then is to come up with a recursive scheme which tries to 'learn' w^* in real time by adaptively updating a running guess as new observations come in.

To arrive at such a scheme, let's pretend to begin with that we do know the underlying law. Assume also that f_w is continuously differentiable in w and let $\nabla^w f_w(\cdot)$ denote its gradient w.r.t. w. The obvious thing to try then is to differentiate the mean square error w.r.t. w and set the derivative equal to zero. Assuming that the interchange of expectation and differentiation is justified, we then have

$$\nabla^w g(w) = -E[\langle Y_n - f_w(X_n), \nabla^w f_w(X_n)\rangle] = 0$$

at the minimum point. We may then seek to minimize the mean square error by gradient descent, given by:

$$
\begin{aligned}
w_{n+1} &= w_n - \nabla^w g(w_n) \\
&= w_n + E[\langle Y_n - f_{w_n}(X_n), \nabla^w f_{w_n}(X_n)\rangle | w_n].
\end{aligned}
$$

This, of course, is not feasible for reasons already mentioned, viz., that the expectation above cannot be evaluated. As a first approximation, we may then consider replacing the expectation by the 'empirical gradient', i.e., the argument of the expectation evaluated at the current guess w_n for w^*,

$$w_{n+1} = w_n + \langle Y_n - f_{w_n}(X_n), \nabla^w f_{w_n}(X_n) \rangle.$$

This, however, will lead to a different kind of problem. The term added to w_n on the right is the nth in a sequence of 'i.i.d. functions' of w, evaluated at w_n. Thus we expect the above scheme to be (and it is) a correlated random walk, zigzagging its way to glory. We may therefore want to smooth it by making only a small, incremental move in the direction suggested by the right-hand side instead of making the full move. This can be achieved by replacing the right-hand side by a convex combination of it and the previous guess w_n, with only a small weight $1 > a(n) > 0$ for the former. That is, we replace the above by

$$w_{n+1} = (1 - a(n))w_n + a(n)(w_n + \langle Y_n - f_{w_n}(X_n), \nabla^w f_{w_n}(X_n) \rangle).$$

Equivalently,

$$w_{n+1} = w_n + a(n)\langle Y_n - f_{w_n}(X_n), \nabla^w f_{w_n}(X_n) \rangle.$$

Once again, if we do not want the scheme to zigzag drastically, we should make $\{a(n)\}$ small, the smaller the better. At the same time, a small $a(n)$ leads to a very small correction to w_n at each iterate, so the scheme will work very slowly, if at all. This suggests starting the iteration with relatively high $\{a(n)\}$ and letting $a(n) \to 0$. (In fact, $a(n) < 1$ as above is not needed, as that can be taken care of by scaling the empirical gradient.) Now let's add and subtract the exact error gradient at the 'known guess' w_n from the empirical gradient on the right-hand side and rewrite the above scheme as

$$
\begin{aligned}
w_{n+1} \;=\; & w_n + a(n)(E[\langle Y_n - f_{w_n}(X_n), \nabla^w f_{w_n}(X_n) \rangle | w_n]) \\[2mm]
& + a(n)(\langle Y_n - f_{w_n}(X_n), \nabla^w f_{w_n}(X_n) \rangle \\[2mm]
& - E[\langle Y_n - f_{w_n}(X_n), \nabla^w f_{w_n}(X_n) \rangle | w_n]).
\end{aligned}
$$

This is of the form

$$w_{n+1} = w_n + a(n)(-\nabla^w g(w_n) + M_{n+1}),$$

with $\{M_n\}$ a martingale difference sequence as in the previous example. One may then view this scheme as a noisy discretization of the o.d.e.

$$\dot{w}(t) = -\nabla^w g(w(t)).$$

This is a particularly well studied o.d.e. We know that it will converge to

$H \overset{\text{def}}{=} \{w : \nabla^w g(w) = 0\}$ in general, and if this set is discrete, to in fact one of the local minima of g for typical (i.e., *generic*: belonging to an open dense set) initial conditions. As before, we are interested in tracking the asymptotic behaviour of this o.d.e. Hence we must ensure that the discrete time steps $\{a(n)\}$ used in the 'noisy discretization' above do cover the entire time axis, i.e.,

$$\sum_n a(n) = \infty, \tag{1.0.1}$$

while retaining $a(n) \to 0$. (Recall from the previous example that $a(n) \to 0$ is needed for asymptotic negligibility of discretization errors.) At the same time, we also want the error due to noise to be asymptotically negligible a.s. The urn example above then suggests that we also impose

$$\sum_n a(n)^2 < \infty, \tag{1.0.2}$$

which asymptotically suppresses the noise variance.

One can show that with (1.0.1) and (1.0.2) in place, for reasonable g (e.g., with $\lim_{\|w\| \to \infty} g(w) = \infty$ and finite H, among other possibilities) the 'stochastic gradient scheme' above will converge a.s. to a local minimum of g.

Once again, what we have here is a special case – perhaps the most important one – of stochastic approximation, analyzed by invoking the 'o.d.e. method'.

What, after all, is stochastic approximation? Historically, stochastic approximation started as a scheme for solving a nonlinear equation $h(x) = 0$ given 'noisy measurements' of the function h. That is, we are given a black box which on input x, gives as its output $h(x) + \xi$, where ξ is a zero mean random variable representing noise. The stochastic approximation scheme proposed by Robbins and Monro (1951)† was to run the iteration

$$x_{n+1} = x_n + a(n)[h(x_n) + M_{n+1}], \tag{1.0.3}$$

where $\{M_n\}$ is the noise sequence and $\{a(n)\}$ are positive scalars satisfying (1.0.1) and (1.0.2) above. The expression in the square brackets on the right is the noisy measurement. That is, $h(x_n)$ and M_{n+1} are not separately available, only their sum is. We shall assume $\{M_n\}$ to be a martingale difference sequence, i.e., a sequence of integrable random variables satisfying

$$E[M_{n+1}|x_m, M_m, m \le n] = 0.$$

This is more general than it appears. For example, an important special case is the d-dimensional iteration

$$x_{n+1} = x_n + a(n)f(x_n, \xi_{n+1}), \ n \ge 0, \tag{1.0.4}$$

† See Lai (2003) for an interesting historical perspective.

for an $f : \mathcal{R}^d \times \mathcal{R}^k \to \mathcal{R}^d$ with i.i.d. noise $\{\xi_n\}$. This can be put in the format of (1.0.3) by defining $h(x) = E[f(x, \xi_1)]$ and $M_{n+1} = f(x_n, \xi_{n+1}) - h(x_n)$ for $n \geq 0$.

Since its inception, the scheme (1.0.3) has been a cornerstone in scientific computation. This has been so largely because of the following advantages, already apparent in the above examples:

- It is designed to handle noisy situations, e.g., the stochastic gradient scheme above. One may say that it captures the average behaviour in the long run. The noise in practice may not only be from measurement errors or approximations, but may also be added deliberately as a probing device or a randomized action, as, e.g., in certain dynamic game situations.
- It is incremental, i.e., it makes small moves at each step. This typically leads to more graceful behaviour of the algorithm at the expense of its speed. We shall say more on this later in the book.
- In typical applications, the computation per iterate is low, making its implementation easy.

These features make the scheme ideal for applications where the key word is 'adaptive'. Thus the stochastic approximation paradigm dominates the fields of adaptive signal processing, adaptive control, and certain subdisciplines of soft computing / artificial intelligence such as neural networks and reinforcement learning – see, e.g., Bertsekas and Tsitsiklis (1997), Haykin (1991) and Haykin (1998). Not surprisingly, it is also emerging as a popular framework for modelling boundedly rational macroeconomic agents – see, e.g., Sargent (1993). The two examples above are representative of these two strands. We shall be seeing many more instances later in this book.

As noted in the preface, there are broadly two approaches to the theoretical analysis of such algorithms. The first, popular with statisticians, is the probabilistic approach based on the theory of martingales and associated objects such as 'almost supermartingales'. The second approach, while still using a considerable amount of martingale theory, views the iteration as a noisy discretization of a limiting o.d.e. Recall that the standard 'Euler scheme' for numerically approximating a trajectory of the o.d.e.

$$\dot{x}(t) = h(x(t))$$

would be

$$x_{n+1} = x_n + ah(x_n),$$

with $x_0 = x(0)$ and $a > 0$ a small time step. The stochastic approximation iteration differs from this in two aspects: replacement of the constant time step 'a' by a time-varying '$a(n)$', and the presence of 'noise' M_{n+1}. This qualifies it as a noisy discretization of the o.d.e. Our aim is to seek x for which $h(x) = 0$, i.e., the equilibrium point(s) of this o.d.e. The o.d.e. would converge (if it does) to these only asymptotically unless it happens to start exactly there. Hence to capture this asymptotic behaviour, we need to track the o.d.e. over the infinite time interval. This calls for the condition $\sum_n a(n) = \infty$. The condition $\sum_n a(n)^2 < \infty$ will on the other hand ensure that the errors due to discretization of the o.d.e. and those due to the noise $\{M_n\}$ both become negligible asymptotically with probability one. (To motivate this, let $\{M_n\}$ be i.i.d. zero mean with a finite variance σ^2. Then by a theorem of Kolmogorov, $\sum_n a(n)M_n$ converges a.s if and only if $\sum_n a(n)^2$ converges.) Together these conditions try to ensure that the iterates do indeed capture the asymptotic behaviour of the o.d.e. We have already seen instances of this above.

Pioneered by Derevitskii and Fradkov (1974), this 'o.d.e. approach' was further extended and introduced to the engineering community by Ljung (1977). It is already the basis of several excellent texts such as Benveniste, Metivier and Priouret (1990), Duflo (1996), and Kushner and Yin (2003), among others†. The rendition here is a slight variation of the traditional one, with an eye on pedagogy so that the highlights of the approach can be introduced quickly and relatively simply. The lecture notes of Benaim (1999) are perhaps the closest in spirit to the treatment here, though at a much more advanced level. (Benaim's notes in particular give an overview of the contributions of Benaim and Hirsch, which introduced important notions from dynamical systems theory, such as internal chain recurrence, to stochastic approximation. These represent a major development in this field in recent years.)

While it is ultimately a matter of personal taste, the o.d.e. approach does indeed appeal to engineers because of the 'dynamical systems' view it takes, which is close to their hearts. Also, as we shall see at the end of this book, it can serve as a useful recipe for concocting new algorithms: any convergent o.d.e. is a potential source of a stochastic approximation algorithm that converges with probability one.

The organization of the book is as follows. Chapter 2 gives the basic convergence analysis for the stochastic approximation algorithm with decreasing stepsizes. This is the core material for the rest of the book. Chapter 3 gives some 'stability tests' that ensure the boundedness of iterates with probability one. Chapter 4 gives some refinements of the results of Chapter 2, viz., an estimate for probability of convergence to a specific attractor if the iterates fall in its domain of attraction. It also gives a result about avoidance with probability one of unstable equilibria. Chapter 5 gives the counterparts of the basic results of Chapter 2 for a more general iteration, which has a differential inclusion as a limit rather than an o.d.e. This is useful in many practical in-

---

† Wasan (1969) and Nevelson and Khasminskii (1976) are two early texts on stochastic approximation, though with a different flavour. See also Ljung et al. (1992).

stances, which are also described in this chapter. Chapter 6 analyzes the cases when more than one timescale is used. This chapter, notably the sections on 'averaging the natural timescale', is technically a little more difficult than the rest and the reader may skip the details of the proofs on a first reading. Chapter 7 describes the distributed asynchronous implementations of the algorithm. Chapter 8 describes the functional central limit theorem for fluctuations associated with the basic scheme of Chapter 2. All the above chapters use decreasing stepsizes. Chapter 9 briefly describes the corresponding theory for constant stepsizes which are popular in some applications.

Chapter 10 of the book has a different flavour: it collects together several examples from engineering, economics, etc., where the stochastic approximation formalism has paid rich dividends. Thus the general techniques of the first part of the book are specialized to each case of interest and the additional structure available in the specific problem under consideration is exploited to say more, depending on the context. It is a mixed bag, the idea being to give the reader a flavour of the various 'tricks of the trade' that may come in handy in future applications. Broadly speaking, one may classify these applications into three strands. The first is the stochastic gradient scheme and its variants wherein h above is either the negative gradient of some function or something close to the negative gradient. This scheme is the underlying paradigm for many adaptive filtering, parameter estimation and stochastic optimization schemes in general. The second is the o.d.e. version of fixed point iterations, i.e., successive application of a map from a space to itself so that it may converge to a point that remains invariant under it (i.e., a fixed point). These are important in a class of applications arising from dynamic programming. The third is the general collection of o.d.e.s modelling collective phenomena in economics etc., such as the urn example above. This classification is, of course, not exhaustive and some instances of stochastic approximation in practice do fall outside of this. Also, we do not consider the continuous time analog of stochastic approximation (see, e.g., Mel'nikov, 1996).

The background required for this book is a good first course on measure theoretic probability, particularly the theory of discrete parameter martingales, at the level of Breiman (1968) or Williams (1991) (though we shall generally refer to Borkar (1995), more out of familiarity than anything), and a first course on ordinary differential equations at the level of Hirsch, Smale and Devaney (2003). There are a few spots where something more than this is required, viz., the theory of weak (Prohorov) convergence of probability measures. The three appendices in Chapter 11 collect together the key aspects of these topics that are needed here.

2
Basic Convergence Analysis

2.1 The o.d.e. limit

In this chapter we begin our formal analysis of the stochastic approximation scheme in $\mathcal{R}^d$ given by

$$x_{n+1} = x_n + a(n)[h(x_n) + M_{n+1}], \ n \geq 0, \qquad (2.1.1)$$

with prescribed x_0 and with the following assumptions which we recall from the last chapter:

(A1) The map $h : \mathcal{R}^d \to \mathcal{R}^d$ is Lipschitz: $\|h(x) - h(y)\| \leq L\|x - y\|$ for some $0 < L < \infty$.

(A2) Stepsizes $\{a(n)\}$ are positive scalars satisfying

$$\sum_n a(n) = \infty, \ \sum_n a(n)^2 < \infty. \qquad (2.1.2)$$

(A3) $\{M_n\}$ is a martingale difference sequence with respect to the increasing family of σ-fields

$$\mathcal{F}_n \overset{\text{def}}{=} \sigma(x_m, M_m, m \leq n) = \sigma(x_0, M_1, \ldots, M_n), \ n \geq 0.$$

That is,

$$E[M_{n+1}|\mathcal{F}_n] = 0 \text{ a.s.}, \ n \geq 0.$$

Furthermore, $\{M_n\}$ are square-integrable with

$$E[\|M_{n+1}\|^2|\mathcal{F}_n] \leq K(1 + \|x_n\|^2) \text{ a.s.}, \ n \geq 0, \qquad (2.1.3)$$

for some constant $K > 0$.

10

Assumption (A1) implies in particular the linear growth condition for $h(\cdot)$:
For a fixed x_0, $||h(x)|| \leq ||h(x_0)|| + L||x - x_0|| \leq K'(1 + ||x||)$ for a suitable
constant $K' > 0$ and all $x \in \mathcal{R}^d$. Thus

$$E[||x_{n+1}||^2]^{\frac{1}{2}} \leq E[||x_n||^2]^{\frac{1}{2}} + a(n)K'(1 + E[||x_n||^2]^{\frac{1}{2}})$$
$$+ a(n)\sqrt{K}(1 + E[||x_n||^2]^{\frac{1}{2}}).$$

We have used here the following fact: $\sqrt{1 + z^2} \leq 1 + z$ for $z \geq 0$. Along
with (2.1.3) and the condition $E[||x_0||^2] < \infty$, this implies inductively that
$E[||x_n||^2], E[||M_n||^2]$ remain bounded for each n.

We shall carry out our analysis under the further assumption:

(A4) The iterates of (2.1.1) remain bounded a.s., i.e.,

$$\sup_n ||x_n|| < \infty, \text{ a.s.} \qquad (2.1.4)$$

This result is far from automatic and usually not very easy to establish.
Some techniques for establishing this will be discussed in the next chapter.

The limiting o.d.e. which (2.1.1) might be expected to track asymptotically
can be written by inspection as

$$\dot{x}(t) = h(x(t)), \ t \geq 0. \qquad (2.1.5)$$

Assumption (A1) ensures that (2.1.5) is well-posed, i.e., has a unique solution for any $x(0)$ that depends continuously on $x(0)$. The basic idea of the
o.d.e. approach to the analysis of (2.1.1) is to construct a suitable continuous interpolated trajectory $\bar{x}(t), t \geq 0$, and show that it asymptotically almost
surely approaches the solution set of (2.1.5). This is done as follows: Define
time instants $t(0) = 0, t(n) = \sum_{m=0}^{n-1} a(m), n \geq 1$. By (2.1.2), $t(n) \uparrow \infty$. Let
$I_n \overset{\text{def}}{=} [t(n), t(n+1)], n \geq 0$. Define a continuous, piecewise linear $\bar{x}(t), t \geq 0$,
by $\bar{x}(t(n)) = x_n, n \geq 0$, with linear interpolation on each interval I_n. That is,

$$\bar{x}(t) = x_n + (x_{n+1} - x_n)\frac{t - t(n)}{t(n+1) - t(n)}, \ t \in I_n.$$

Note that $\sup_{t \geq 0} ||\bar{x}(t)|| = \sup_n ||x_n|| < \infty$ a.s. Let $x^s(t), t \geq s$, denote the
unique solution to (2.1.5) 'starting at s':

$$\dot{x}^s(t) = h(x^s(t)), \ t \geq s,$$

with $x^s(s) = \bar{x}(s), s \in \mathcal{R}$. Likewise, let $x_s(t), t \leq s$, denote the unique solution
to (2.1.5) 'ending at s':

$$\dot{x}_s(t) = h(x_s(t)), \ t \leq s,$$

with $x_s(s) = \bar{x}(s)$, $s \in \mathcal{R}$. Define also

$$\zeta_n = \sum_{m=0}^{n-1} a(m)M_{m+1}, \quad n \geq 1.$$

By (A3) and the remarks that follow, $(\zeta_n, \mathcal{F}_n), n \geq 1$, is a zero mean, square-integrable martingale. Furthermore, by (A2), (A3) and (A4),

$$\sum_{n \geq 0} E[\|\zeta_{n+1} - \zeta_n\|^2 | \mathcal{F}_n] = \sum_{n \geq 0} a(n)^2 E[\|M_{n+1}\|^2 | \mathcal{F}_n] < \infty, \quad \text{a.s.}$$

It follows from the martingale convergence theorem (Appendix C) that ζ_n converges a.s. as $n \to \infty$.

Lemma 1. *For any $T > 0$,*

$$\lim_{s \to \infty} \sup_{t \in [s, s+T]} \|\bar{x}(t) - x^s(t)\| = 0, \quad a.s.$$

$$\lim_{s \to \infty} \sup_{t \in [s-T, s]} \|\bar{x}(t) - x_s(t)\| = 0, \quad a.s.$$

Proof. We shall only prove the first claim, as the arguments for proving the second claim are completely analogous. Let $t(n+m)$ be in $[t(n), t(n)+T]$. Let $[t] \stackrel{\text{def}}{=} \max\{t(k) : t(k) \leq t\}$. Then by construction,

$$\bar{x}(t(n+m)) = \bar{x}(t(n)) + \sum_{k=0}^{m-1} a(n+k)h(\bar{x}(t(n+k))) + \delta_{n,n+m}, \quad (2.1.6)$$

where $\delta_{n,n+m} \stackrel{\text{def}}{=} \zeta_{n+m} - \zeta_n$. Compare this with

$$\begin{aligned}
x^{t(n)}(t(m+n)) &= \bar{x}(t(n)) + \int_{t(n)}^{t(n+m)} h(x^{t(n)}(t))dt \\
&= \bar{x}(t(n)) + \sum_{k=0}^{m-1} a(n+k)h(x^{t(n)}(t(n+k))) \\
&\quad + \int_{t(n)}^{t(n+m)} (h(x^{t(n)}(y)) - h(x^{t(n)}([y])))dy.
\end{aligned}$$

$$(2.1.7)$$

We shall now bound the integral on the right-hand side. Let $C_0 \stackrel{\text{def}}{=} \sup_n \|x_n\| < \infty$ a.s., let $L > 0$ denote the Lipschitz constant of h as before, and let $s \leq t \leq s + T$. Note that $\|h(x) - h(0)\| \leq L\|x\|$, and so $\|h(x)\| \leq \|h(0)\| + L\|x\|$. Since

$x^s(t) = \bar{x}(s) + \int_s^t h(x^s(\tau))d\tau,$

$$\|x^s(t)\| \leq \|\bar{x}(s)\| + \int_s^t [\|h(0)\| + L\|x^s(\tau)\|]d\tau$$

$$\leq (C_0 + \|h(0)\|T) + L\int_s^t \|x^s(\tau)\|d\tau.$$

By Gronwall's inequality (see Appendix B), it follows that

$$\|x^s(t)\| \leq (C_0 + \|h(0)\|T)e^{LT}, \quad s \leq t \leq s + T.$$

Thus, for all $s \leq t \leq s + T$,

$$\|h(x^s(t))\| \leq C_T \stackrel{\text{def}}{=} \|h(0)\| + L(C_0 + \|h(0)\|T)e^{LT} < \infty, \quad \text{a.s.}$$

Now, if $0 \leq k \leq (m-1)$ and $t \in (t(n+k), t(n+k+1)]$,

$$\|x^{t(n)}(t) - x^{t(n)}(t(n+k))\| \leq \left\| \int_{t(n+k)}^t h(x^{t(n)}(s))ds \right\|$$

$$\leq C_T(t - t(n+k))$$

$$\leq C_T a(n+k).$$

Thus,

$$\left\| \int_{t(n)}^{t(n+m)} (h(x^{t(n)}(t)) - h(x^{t(n)}([t])))dt \right\|$$

$$\leq \int_{t(n)}^{t(n+m)} L\|x^{t(n)}(t) - x^{t(n)}([t])\|dt$$

$$= L\sum_{k=0}^{m-1} \int_{t(n+k)}^{t(n+k+1)} \|x^{t(n)}(t) - x^{t(n)}(t(n+k))\|dt$$

$$\leq C_T L\sum_{k=0}^{m-1} a(n+k)^2$$

$$\leq C_T L\sum_{k=0}^{\infty} a(n+k)^2 \stackrel{n\uparrow\infty}{\to} 0, \quad \text{a.s.} \tag{2.1.8}$$

Also, since the martingale $(\zeta_n, \mathcal{F}_n)$ converges a.s., we have

$$\sup_{k\geq 0} \|\delta_{n,n+k}\| \stackrel{n\uparrow\infty}{\to} 0, \quad \text{a.s.} \tag{2.1.9}$$

Subtracting (2.1.7) from (2.1.6) and taking norms, we have

$$\|\bar{x}(t(n+m)) - x^{t(n)}(t(n+m))\|$$

$$\leq L \sum_{i=0}^{m-1} a(n+i)\|\bar{x}(t(n+i)) - x^{t(n)}(t(n+i))\|$$

$$+ C_T L \sum_{k\geq 0} a(n+k)^2 + \sup_{k\geq 0} \|\delta_{n,n+k}\|, \quad \text{a.s.}$$

Define $K_{T,n} = C_T L \sum_{k\geq 0} a(n+k)^2 + \sup_{k\geq 0}\|\delta_{n,n+k}\|$. Note that $K_{T.n} \to 0$ a.s. as $n \to \infty$. Also, let $z_i = \|\bar{x}(t(n+i)) - x^{t(n)}(t(n+i))\|$ and $b_i \stackrel{def}{=} a(n+i)$. Thus, the above inequality becomes

$$z_m \leq K_{T,n} + L \sum_{i=0}^{m-1} b_i z_i.$$

Note that $z_0 = 0$ and $\sum_{i=0}^{m-1} b_i \leq T$. The discrete Gronwall lemma (see Appendix B) tells us that

$$\sup_{0\leq i\leq m} z_i \leq K_{T,n} e^{LT}.$$

One then has that for $t(n+m) \leq t(n) + T$,

$$\|\bar{x}(t(n+m)) - x^{t(n)}(t(n+m))\| \leq K_{T,n} e^{LT}, \quad \text{a.s.}$$

If $t(n+k) \leq t \leq t(n+k+1)$, we have that

$$\bar{x}(t) = \lambda \bar{x}(t(n+k)) + (1-\lambda)\bar{x}(t(n+k+1))$$

for some $\lambda \in [0,1]$. Thus,

$$\|x^{t(n)}(t) - \bar{x}(t)\|$$

$$= \|\lambda(x^{t(n)}(t) - \bar{x}(t(n+k))) + (1-\lambda)(x^{t(n)}(t) - \bar{x}(t(n+k+1)))\|$$

$$\leq \lambda\|x^{t(n)}(t(n+k)) - \bar{x}(t(n+k)) + \int_{t(n+k)}^{t} h(x^{t(n)}(s))ds\|$$

$$+ (1-\lambda)\|x^{t(n)}(t(n+k+1)) - \bar{x}(t(n+k+1))$$

$$- \int_{t}^{t(n+k+1)} h(x^{t(n)}(s))ds\|$$

$$\leq (1-\lambda)\|x^{t(n)}(t(n+k+1)) - \bar{x}(t(n+k+1))\|$$

$$+ \lambda\|x^{t(n)}(t(n+k)) - \bar{x}(t(n+k))\|$$

$$+ \max(\lambda, 1-\lambda) \int_{t(n+k)}^{t(n+k+1)} \|h(x^{t(n)}(s))\|ds.$$

Since $\|h(x^s(t))\| \leq C_T$ for all $s \leq t \leq s + T$, it follows that

$$\sup_{t\in[t(n),t(n)+T]} \|\bar{x}(t) - x^{t(n)}(t)\| \leq K_{T,n} e^{LT} + C_T a(n+k), \quad \text{a.s.}$$

The claim now follows for the special case of $s \to \infty$ along $\{t(n)\}$. The general claim follows easily from this special case. ∎

Recall that a closed set $A \subset \mathcal{R}^d$ is said to be an *invariant set* (resp. a positively / negatively invariant set) for the o.d.e. (2.1.5) if any trajectory $x(t), -\infty < t < \infty$ (resp. $0 \leq t < \infty / -\infty < t \leq 0$) of (2.1.5) with $x(0) \in A$ satisfies $x(t) \in A \: \forall t \in \mathcal{R}$ (resp. $\forall t \geq 0 / \forall t \leq 0$). It is said to be *internally chain transitive* in addition if for any $x, y \in A$ and any $\epsilon > 0, T > 0$, there exist $n \geq 1$ and points $x_0 = x, x_1, \ldots, x_{n-1}, x_n = y$ in A such that the trajectory of (2.1.5) initiated at x_i meets with the ϵ-neighbourhood of x_{i+1} for $0 \leq i < n$ after a time $\geq T$. (If we restrict to $y = x$ in the above, the set is said to be *internally chain recurrent*.) Let $\Phi_t : \mathcal{R}^d \to \mathcal{R}^d$ denote the map that takes $x(0)$ to $x(t)$ via (2.1.5). Under our conditions on h, this map will be continuous (in fact Lipschitz) for each $t > 0$. (See Appendix B.) From the uniqueness of solutions to (2.1.5) in both forward and backward time, it follows that Φ_t is invertible. In fact it turns out to be a homeomorphism, i.e., a continuous bijection with a continuous inverse (see Appendix B). Thus we can define $\Phi_{-t}(x) = \Phi_t^{-1}(x)$, the point at which the trajectory starting at time 0 at x and running backward in time for a duration t would end up. Along with $\Phi_0 \equiv$ the identity map on $\mathcal{R}^d, \{\Phi_t, t \in \mathcal{R}\}$ defines a group of homeomorphisms on $\mathcal{R}^d$, which is referred to as the *flow* associated with (2.1.5). Thus the definition of an invariant set can be recast as follows: A is invariant if

$$\Phi_t(A) = A \quad \forall t \in \mathcal{R}.$$

A corresponding statement applies to positively or negatively invariant sets with $t \geq 0$, resp. $t \leq 0$. Our general convergence theorem for stochastic approximation, due to Benaim (1996), is the following.

Theorem 2. *Almost surely, the sequence $\{x_n\}$ generated by (2.1.1) converges to a (possibly sample path dependent) compact connected internally chain transitive invariant set of (2.1.5).*

Proof. Consider a sample point where (2.1.4) and the conclusions of Lemma 1 hold. Let A denote the set $\bigcap_{t \geq 0} \overline{\{\bar{x}(s) : s \geq t\}}$. Since $\bar{x}(\cdot)$ is continuous and bounded, $\overline{\{\bar{x}(s) : s \geq t\}}, t \geq 0$, is a nested family of nonempty compact and connected sets. A, being the intersection thereof, will also be nonempty compact and connected. Then $\bar{x}(t) \to A$ and therefore $x_n \to A$. In fact, for any $\epsilon > 0$, let $A^\epsilon \stackrel{\text{def}}{=} \{x : \min_{y \in A} \|x - y\| < \epsilon\}$. Then $(A^\epsilon)^c \cap (\bigcap_{t \geq 0} \overline{\{\bar{x}(s) : s \geq t\}}) = \phi$. Hence by the finite intersection property of families of compact sets, $(A^\epsilon)^c \cap \overline{\{\bar{x}(s) : s \geq t'\}} = \emptyset$ for some $t' > 0$. That is, $\bar{x}(t' + \cdot) \in A^\epsilon$. Conversely, if $x \in A$, there exist $s_n \uparrow \infty$ in $[0, \infty)$ such that $\bar{x}(s_n) \to x$. This is immediate from the definition of A. In fact, we have

$$\max_{s\in[t(n),t(n+1)]} \|\bar{x}(s) - \bar{x}(t(n))\| = O(a(n)) \to 0$$

as $n \to \infty$. Thus we may take $s_n = t(m(n))$ for suitable $\{m(n)\}$ without any loss of generality. Let $\tilde{x}(\cdot)$ denote the trajectory of (2.1.5) with $\tilde{x}(0) = x$. Then by the first part of Lemma 1 and the continuity of the map Φ_t defined above, it follows that $x^{s_n}(s_n + t) = \Phi_t(\bar{x}(s_n)) \to \Phi_t(x) = \tilde{x}(t)$ for all $t > 0$. By Lemma 1, $\bar{x}(s_n + t) \to \tilde{x}(t)$, implying that $\tilde{x}(t) \in A$ as well. A similar argument works for $t < 0$, using the second part of Lemma 1. Thus A is invariant under (2.1.5).

Let $\tilde{x}_1, \tilde{x}_2 \in A$ and fix $\epsilon > 0, T > 0$. Pick $\epsilon/4 > \delta > 0$ such that: if $\|z-y\| < \delta$ and $\hat{x}_z(\cdot), \hat{x}_y(\cdot)$ are solutions to (2.1.5) with initial conditions z, y resp., then $\max_{t\in[0,2T]} \|\hat{x}_z(t) - \hat{x}_y(t)\| < \epsilon/4$. Also pick $n_0 > 1$ such that $s \geq t(n_0)$ implies that $\bar{x}(s + \cdot) \in A^\delta$ and $\sup_{t\in[s,s+2T]} \|\bar{x}(t) - x^s(t)\| < \delta$. Pick $n_2 > n_1 \geq n_0$ such that $\|\bar{x}(t(n_i)) - \tilde{x}_i\| < \delta, i = 1, 2$. Let $kT \leq t(n_2) - t(n_1) < (k+1)T$ for some integer $k \geq 0$ and let $s(0) = t(n_1), s(i) = s(0) + iT$ for $1 \leq i < k$, and $s(k) = t(n_2)$. Then for $0 \leq i < k$, $\sup_{t\in[s(i),s(i+1)]} \|\bar{x}(t) - x^{s(i)}(t)\| < \delta$. Pick $\hat{x}_i, 0 \leq i \leq k$, in A such that $\hat{x}_1 = \tilde{x}_1$, $\hat{x}_k = \tilde{x}_2$, and for $0 < i < k$, $\hat{x}_i$ are in the δ-neighbourhood of $\bar{x}(s(i))$. The sequence $(s(i), \hat{x}_i), 0 \leq i \leq k$, satisfies the definition of internal chain transitivity: If $x_i^*(\cdot)$ denotes the trajectories of (2.1.5) initiated at $\hat{x}_i$ for each i, we have

$$\|x_i^*(s(i+1) - s(i)) - \hat{x}_{i+1}\|$$
$$\leq \quad \|x_i^*(s(i+1) - s(i)) - x^{s(i)}(s(i+1))\|$$
$$\quad + \|x^{s(i)}(s(i+1)) - \bar{x}(s(i+1))\| + \|\bar{x}(s(i+1)) - \hat{x}_{i+1}\|$$
$$\leq \quad \frac{\epsilon}{4} + \frac{\epsilon}{4} + \frac{\epsilon}{4} < \epsilon.$$

This completes the proof.   ∎

2.2 Extensions and variations

Some important extensions of the foregoing are immediate:

- When the set $\{\sup_n \|x_n\| < \infty\}$ has a positive probability not necessarily equal to one, we still have

$$\sum_n a(n)E[\|M_{n+1}\|^2|\mathcal{F}_n] < \infty$$

a.s. *on this set.* The martingale convergence theorem from Appendix C cited in the proof of Lemma 1 above then tells us that ζ_n converges a.s. on this set. Thus by the same arguments as before (which are *pathwise*), Theorem 2 continues to hold 'a.s. on the set $\{\sup_n \|x_n\| < \infty\}$'.

- While we took $\{a(n)\}$ to be deterministic in section 2.1, the arguments would also go through if $\{a(n)\}$ are random and bounded, satisfy (A2) with probability one, and (A3) holds, with $\mathcal{F}_n$ redefined as

$$\mathcal{F}_n = \sigma(x_m, M_m, a(m), m \leq n)$$

for $n \geq 0$. In fact, the boundedness condition for random $\{a(n)\}$ could be relaxed by imposing appropriate moment conditions. We shall not get into the details of this at any point, but it is worth keeping in mind throughout as there are applications (e.g., in system identification) when $\{a(n)\}$ are random.

- The arguments above go through even if we replace (2.1.1) by

$$x_{n+1} = x_n + a(n)[h(x_n) + M_{n+1} + \epsilon(n)], \; n \geq 0,$$

where $\{\epsilon(n)\}$ is a deterministic or random bounded sequence which is $o(1)$. This is because $\{\epsilon(n)\}$ then contributes an additional error term in the proof of Lemma 1 which is also asymptotically negligible and therefore does not affect the conclusions. This important observation will be recalled often in what follows.

These observations apply throughout the book wherever the arguments are pathwise, i.e., except in Chapter 9. The next corollary is often useful in narrowing down the potential candidates for A.

Suppose there exists a continuously differentiable $V : \mathcal{R}^d \to [0, \infty)$ such that $\lim_{\|x\| \to \infty} V(x) = \infty$, $H \stackrel{\text{def}}{=} \{x \in \mathcal{R}^d : V(x) = 0\} \neq \phi$, and $\langle h(x), \nabla V(x) \rangle \leq 0$ with equality if and only if $x \in H$. (Thus V is a 'Liapunov function'.)

Corollary 3. *Almost surely, $\{x_n\}$ converge to an internally chain transitive invariant set contained in H.*

Proof. The argument is sample pathwise for a sample path in the probability one set where assumption (A4) and Lemma 1 hold. Fix one such sample path and let $C' = \sup_n \|x_n\|$ and $C = \sup_{\|x\| \leq C'} V(x)$. For any $0 < a \leq C$, let $H^a \stackrel{\text{def}}{=} \{x \in \mathcal{R}^d : V(x) < a\}$, and let $\bar{H}^a$ denote the closure of H^a. Fix an η such that $0 < \eta < C/2$. Let

$$\Delta \stackrel{\text{def}}{=} \min_{x \in \bar{H}^C \setminus H^\eta} |\langle h(x), \nabla V(x) \rangle| > 0.$$

Let T be an upper bound for the time required for a solution of $\dot{x} = h(x)$ to reach H^η, starting from a point in H^C. We may choose $T > C/\Delta$. Let $\delta > 0$ be such that for $x \in \bar{H}^C$ and $\|x - y\| < \delta$, we have $|V(x) - V(y)| < \eta$. Such a choice of δ is possible by the uniform continuity of V on compact sets. By Lemma 1, there is a t_0 such that for all $t \geq t_0$, $\sup_{s \in [t, t+T]} \|\bar{x}(s) - x^t(s)\| < \delta$. Note that $\bar{x}(\cdot) \in \bar{H}^C$, and so for all $t \geq t_0$, $|V(\bar{x}(t+T)) - V(x^t(t+T))| < \eta$.

But $x^t(t+T) \in H^\eta$ and therefore $\bar{x}(t+T) \in H^{2\eta}$. Thus for all $t \geq t_0 + T$, $\bar{x}(t) \in H^{2\eta}$. Since η can be taken to be arbitrarily small, it follows that $\bar{x}(t) \to H$ as $t \to \infty$.   ∎

Alternatively, we can invoke the 'LaSalle invariance principle' (see Appendix B) in conjunction with Theorem 2. The following corollary is immediate:

Corollary 4. *If the only internally chain transitive invariant sets for (2.1.5) are isolated equilibrium points, then $\{x_n\}$ a.s. converges to a possibly sample path dependent equilibrium point.*

More generally, a similar statement could be made for isolated internally chain transitive invariant sets, i.e., internally chain transitive invariant sets each of which is at a strictly positive distance from the rest. We shall refine Corollary 4 in Chapter 4. The next corollary is a variation of the so-called 'Kushner–Clark lemma' (Kushner and Clark, 1978) and also follows from the above discussion. Recall that 'i.o.' in probability theory stands for 'infinitely often'.

Corollary 5. *Let G be an open set containing a bounded internally chain transitive invariant set D for (2.1.5), and suppose that $\bar{G}$ does not intersect any other bounded internally chain transitive invariant set (except possibly subsets of D). Then under (A4), $x_n \to D$ a.s. on the set $\{x_n \in G \ i.o.\} \stackrel{def}{=} \bigcap_n \bigcup_{m \geq n} \{x_m \in G\}$.*

Proof. We know that a.s., x_n converges to a compact, connected internally chain transitive invariant set. Let this set be D'. If D' does not intersect $\bar{G}$, then by compactness of D', there is an ϵ-neighbourhood $N_\epsilon(D')$ of D' which does not intersect $\bar{G}$. But since $x_n \to D'$, $x_n \in N_\epsilon(D')$ for n large. This, however, leads to a contradiction if $x_n \in G$ i.o. Thus, if $x_n \in G$ i.o., D' has to intersect $\bar{G}$. It follows that D' equals D or a subset thereof, and so $x_n \to D$ a.s. on the set $\{x_n \in G \ i.o.\}$.   ∎

In the more general set-up of Theorem 2, the next theorem is sometimes useful. (The statement and proof require some familiarity with weak (Prohorov) convergence of probability measures. See Appendix C for a brief account.)

Let $\mathcal{P}(\mathcal{R}^d)$ denote the space of probability measures on $\mathcal{R}^d$ with Prohorov topology (also known as the topology of weak convergence, see, e.g., Borkar, 1995, Chapter 2). Let $C_0(\mathcal{R}^d)$ denote the space of continuous functions on $\mathcal{R}^d$ that vanish at infinity. Then the space M of complex Borel measures on $\mathcal{R}^d$ is isomorphic to the dual space $C_0^*(\mathcal{R}^d)$. The isomorphism is given by $\mu \mapsto \int (\cdot) d\mu$. (See, e.g., Rudin, 1986, Chapter 6.) It is easy to show that $\mathcal{P}(\mathcal{R}^d)$ consists of real measures μ which correspond to those elements $\hat{\mu}$ of $C_0^*(\mathcal{R}^d)$ that are nonnegative on nonnegative functions in $C_0(\mathcal{R}^d)$ (i.e., $f \geq 0$

for $f \in C_0(\mathcal{R}^d)$ implies that $\hat{\mu}(f) \geq 0$), *and for a constant function* $f(\cdot) \equiv C$, $\hat{\mu}(f) = C$.

Define (random) measures $\nu(t), t > 0$, on $\mathcal{R}^d$ by

$$\int f d\nu(t) = \frac{1}{t} \int_0^t f(\bar{x}(s)) ds$$

for $f \in C_0(\mathcal{R}^d)$. These are called *empirical measures*. Since this integral is nonnegative for nonnegative f and furthermore, $\nu(t)(\mathcal{R}^d) = \frac{1}{t} \int_0^t 1\, ds = 1$, $\nu(t)$ is a probability measure on $\mathcal{R}^d$. By (A4), the $\nu(t)$ are supported in a common compact subset of $\mathcal{R}^d$ independent of t. By Prohorov's theorem (see Appendix C), they form a relatively compact subset of $\mathcal{P}(\mathcal{R}^d)$.

Theorem 6. *Almost surely, every limit point ν^* of $\nu(t)$ in $\mathcal{P}(\mathcal{R}^d)$ as $t \to \infty$ is invariant under (2.1.5).*

Proof. Fix some $s > 0$. Consider a sample path for which Lemma 1 applies, i.e., for which $\|\bar{x}(y + s) - x^y(y + s)\| \to 0$ as $y \to \infty$. Let $f \in C_0(\mathcal{R}^d)$. Note that

$$\left| \frac{1}{t} \int_0^t f(\bar{x}(y)) dy - \frac{1}{t} \int_s^{t+s} f(\bar{x}(y)) dy \right| \to 0 \text{ as } t \to \infty.$$

Note also that the quantity on the left above is the same as

$$\left| \frac{1}{t} \int_0^t f(\bar{x}(y)) dy - \frac{1}{t} \int_0^t f(\bar{x}(y + s)) dy \right|.$$

Let $\epsilon > 0$. By uniform continuity of f, there is a T such that for $y \geq T$, $\|f(\bar{x}(y + s)) - f(x^y(y + s))\| < \epsilon$. Now, if $t \geq T$,

$$\left| \frac{1}{t} \int_0^t f(\bar{x}(y + s)) dy - \frac{1}{t} \int_0^t f(x^y(y + s)) dy \right|$$

$$\leq \frac{1}{t} \int_0^T |f(\bar{x}(y + s)) - f(x^y(y + s))| dy$$

$$+ \frac{1}{t} \int_T^t |f(\bar{x}(y + s)) - f(x^y(y + s))| dy$$

$$\leq \frac{T}{t} 2B + \frac{(t - T)}{t} \epsilon \leq 2\epsilon$$

for t large enough. Here B is a bound on the magnitude of $f \in C_0(\mathcal{R}^d)$. Thus

$$\left| \frac{1}{t} \int_0^t f(\bar{x}(y + s)) dy - \frac{1}{t} \int_0^t f(x^y(y + s)) dy \right| \to 0 \text{ as } t \to \infty.$$

But since

$$\left| \frac{1}{t} \int_0^t f(\bar{x}(y)) dy - \frac{1}{t} \int_0^t f(\bar{x}(y + s)) dy \right| \to 0$$

as $t \to \infty$, it follows that

$$|\frac{1}{t} \int_0^t f(\bar{x}(y))dy - \frac{1}{t} \int_0^t f(x^y(y+s))dy| \to 0$$

as $t \to \infty$. But this implies that

$$|\int f d\nu(t) - \int f \circ \Phi_s d\nu(t)| = |\frac{1}{t} \int_0^t f(\bar{x}(y))dy - \frac{1}{t} \int_0^t f \circ \Phi_s(\bar{x}(y))dy|$$

$$= |\frac{1}{t} \int_0^t f(\bar{x}(y))dy - \frac{1}{t} \int_0^t f(x^y(y+s))dy|$$

$$\to 0 \text{ as } t \to \infty.$$

If ν^* is a limit point of $\nu(t)$, there is a sequence $t_n \nearrow \infty$ such that $\nu(t_n) \to \nu^*$ weakly. Thus, $\int f d\nu(t_n) \to \int f d\nu^*$ and $\int f \circ \Phi_s d\nu(t_n) \to \int f \circ \Phi_s d\nu^*$. But $|\int f d\nu(t_n) - \int f \circ \Phi_s d\nu(t_n)| \to 0$ as $n \to \infty$. This tells us that $\int f d\nu^* = \int f \circ \Phi_s d\nu^*$. This holds for all $f \in C_0(\mathcal{R}^d)$. Hence ν^* is invariant under Φ_s. As $s > 0$ was arbitrary, the claim follows.   ■

See Benaim and Schreiber (2001) for further results in this vein.

We conclude this section with some comments regarding stepsize selection. Our view of $\{a(n)\}$ as discrete time steps in the o.d.e. approximation already gives some intuition about their role. Thus large stepsizes will mean faster simulation of the o.d.e., but also larger errors due to discretization and noise (the latter is so because the stepsize $a(n)$ also multiplies the 'noise' M_{n+1} in the algorithm). Reducing the stepsizes would mean lower discretization errors and noise-induced errors and therefore a more graceful behaviour of the algorithm, but at the expense of a slower speed of convergence. This is because one is taking a larger number of iterations to simulate any given time interval in 'o.d.e. time'. In the parlance of artificial intelligence, larger stepsizes aid better *exploration* of the solution space, while smaller stepsizes aid better *exploitation* of the local information available. The trade-off between them is a well-known rule of thumb in AI. Starting with a relatively large $a(n)$ and decreasing it slowly tries to strike a balance between the two. See Goldstein (1988) for some results on stepsize selection.

3
Stability Criteria

3.1 Introduction

In this chapter we discuss a couple of schemes for establishing the a.s. boundedness of iterates assumed above. The convergence analysis of the preceding chapter has some universal applicability, but the situation is different for stability criteria. There are several variations of stability criteria applicable under specific restrictions and sometimes motivated by specific applications for which they are tailor-made. (The second test we see below is one such.) We describe only a couple of these variations. The first one is quite broadly applicable. The second is a bit more specialized, but has been included because it has a distinct flavour and shows how one may tweak known techniques such as stochastic Liapunov functions to obtain new and useful criteria.

3.2 Stability criterion

The first scheme is adapted from Borkar and Meyn (2000). The idea of this test is as follows: We consider the piecewise linear interpolated trajectory $\bar{x}(\cdot)$ at times $T_n \uparrow \infty$, which are spaced approximately $T > 0$ apart and divide the time axis into concatenated time segments of length approximately T. If at any T_n, the iterate has gone out of the unit ball in $\mathcal{R}^d$, we rescale it over the segment $[T_n, T_{n+1})$ by dividing it by the norm of its value at T_n. If the original trajectory drifts towards infinity, then there is a corresponding sequence of rescaled segments as above that asymptotically track a limiting o.d.e. obtained as a scaling limit of our 'basic o.d.e.'

$$\dot{x}(t) = h(x(t)). \tag{3.2.1}$$

If this scaling limit is globally asymptotically stable to the origin, these segments, and therefore the original iterations which differ from them only by a scaling factor, should start drifting towards the origin, implying stability.

Formally, assume the following:

(A5) The functions $h_c(x) \overset{\text{def}}{=} h(cx)/c, c \geq 1, x \in \mathcal{R}^d$, satisfy $h_c(x) \rightarrow h_\infty(x)$ as $c \rightarrow \infty$, uniformly on compacts for some $h_\infty \in C(\mathcal{R}^d)$. Furthermore, the o.d.e.

$$\dot{x}(t) = h_\infty(x(t)) \tag{3.2.2}$$

has the origin as its unique globally asymptotically stable equilibrium.

The o.d.e. (3.2.2) is the aforementioned 'scaling limit'. It is worth noting here that

(i) h_c, h_∞ will be Lipschitz with the same Lipschitz constant as h, implying in particular the well-posedness of (3.2.2) above and also of the o.d.e.

$$\dot{x}(t) = h_c(x(t)). \tag{3.2.3}$$

In particular, they are equicontinuous. Thus pointwise convergence of h_c to h_∞ as $c \rightarrow \infty$ will automatically imply uniform convergence on compacts.

(ii) h_∞ satisfies $h_\infty(ax) = ah_\infty(x)$ for $a > 0$, and hence if (3.2.2) has an isolated equilibrium, it must be at the origin.

(iii) $\|h_c(x) - h_c(0)\| \leq L\|x\|$, and so $\|h_c(x)\| \leq \|h_c(0)\| + L\|x\| \leq \|h(0)\| + L\|x\| \leq K_0(1 + \|x\|)$ (for a suitable constant K_0).

Let $\phi_\infty(t, x)$ denote the solution of the o.d.e. $\dot{x} = h_\infty(x)$ with initial condition x.

Lemma 1. *There exists a $T > 0$ such that for all initial conditions x on the unit sphere, $\|\phi_\infty(t, x)\| < \frac{1}{8}$ for all $t > T$.*

Proof. Since asymptotic stability implies Liapunov stability (see Appendix B), there is a $\delta > 0$ such that any trajectory starting within distance δ of the origin stays within distance $\frac{1}{8}$ thereof. For an initial condition x on the unit sphere, let T_x be a time at which the solution is within distance $\delta/2$ of the origin. Let y be any other initial condition on the unit sphere. Note that

$$\phi_\infty(t, x) = x + \int_0^t h_\infty(\phi_\infty(s, x))ds, \text{ and}$$

$$\phi_\infty(t, y) = y + \int_0^t h_\infty(\phi_\infty(s, y))ds.$$

Subtracting the above equations and using the Lipschitz property, we get

$$\|\phi_\infty(t,x) - \phi_\infty(t,y)\| \leq \|x - y\| + L \int_0^t \|\phi_\infty(s,x) - \phi_\infty(s,y)\| ds.$$

Then by Gronwall's inequality we find that for $t \leq T_x$,

$$\|\phi_\infty(t,x) - \phi_\infty(t,y)\| \leq \|x - y\| e^{LT_x}.$$

So there is a neighbourhood U_x of x such that for all $y \in U_x$, $\phi_\infty(T_x, y)$ is within distance δ of the origin. By Liapunov stability, this implies that $\phi_\infty(t, y)$ remains within distance $\frac{1}{8}$ of the origin for all $t \geq T_x$. Since the unit sphere is compact, it can be covered by a finite number of such neighbourhoods $U_{x_1}, \ldots, U_{x_n}$ with corresponding times $T_{x_1}, \ldots, T_{x_n}$. Then the statement of the lemma holds if T is the maximum of $T_{x_1}, \ldots, T_{x_n}$. ∎

The following lemma shows that the solutions of the o.d.e.s $\dot{x} = h_c(x)$ and $\dot{x} = h_\infty(x)$ are close for c large enough.

Lemma 2. *Let $K \subset \mathcal{R}^d$ be compact, and let $[0, T]$ be a given time interval. Then for $t \in [0, T]$ and $x_0 \in K$,*

$$\|\phi_c(t, x) - \phi_\infty(t, x_0)\| \leq [\|x - x_0\| + \epsilon(c)T]e^{LT},$$

where $\epsilon(c)$ is independent of $x_0 \in K$ and $\epsilon(c) \to 0$ as $c \to \infty$. In particular, if $x = x_0$, then

$$\|\phi_c(t, x_0) - \phi_\infty(t, x_0)\| \leq \epsilon(c)Te^{LT}. \tag{3.2.4}$$

Proof. Note that

$$\phi_c(t, x) = x + \int_0^t h_c(\phi_c(s, x))ds, \text{ and}$$

$$\phi_\infty(t, x_0) = x_0 + \int_0^t h_\infty(\phi_\infty(s, x_0))ds.$$

This gives

$$\|\phi_c(t, x) - \phi_\infty(t, x_0)\| \leq \|x - x_0\| + \int_0^t \|h_c(\phi_c(s, x)) - h_\infty(\phi_\infty(s, x_0))\| ds.$$

Now, using the facts that $\phi_\infty([0, T], K)$ is compact, $h_c \to h_\infty$ uniformly on compact sets, and h_c has the Lipschitz property, we get

$$\|h_c(\phi_c(s, x)) - h_\infty(\phi_\infty(s, x_0))\|$$
$$\leq \|h_c(\phi_c(s, x)) - h_c(\phi_\infty(s, x_0))\|$$
$$+ \|h_c(\phi_\infty(s, x_0)) - h_\infty(\phi_\infty(s, x_0))\|$$
$$\leq L\|\phi_c(s, x) - \phi_\infty(s, x_0)\| + \epsilon(c),$$

where $\epsilon(c)$ is independent of $x_0 \in K$ and $\epsilon(c) \to 0$ as $c \to \infty$. Thus for $t \leq T$, we get

$$\|\phi_c(t,x) - \phi_\infty(t,x_0)\| \leq \|x - x_0\| + \epsilon(c)T + L \int_0^t \|\phi_c(s,x) - \phi_\infty(s,x_0)\| ds.$$

The conclusion follows from Gronwall's inequality. ∎

The previous two lemmas give us:

Corollary 3. *There exist $c_0 > 0$ and $T > 0$ such that for all initial conditions x on the unit sphere, $\|\phi_c(t,x)\| < \frac{1}{4}$ for $t \in [T, T+1]$ and $c > c_0$.*

Proof. Choose T as in Lemma 1. Now, using equation (3.2.4) with K taken to be the closed unit ball, conclude that $\|\phi_c(t,x)\| < \frac{1}{4}$ for $t \in [T, T+1]$ and c such that $\epsilon(c)(T+1)e^{L(T+1)} < \frac{1}{8}$. ∎

Let $T_0 = 0$ and $T_{n+1} = \min\{t(m) : t(m) \geq T_n + T\}$ for $n \geq 0$. Then $T_{n+1} \in [T_n + T, T_n + T + \bar{a}]$ $\forall n$, where $\bar{a} = \sup_n a(n)$, $T_n = t(m(n))$ for suitable $m(n) \uparrow \infty$, and $T_n \uparrow \infty$. For notational simplicity, let $\bar{a} = 1$ without loss of generality. Define the piecewise continuous trajectory $\hat{x}(t), t \geq 0$, by $\hat{x}(t) = \bar{x}(t)/r(n)$ for $t \in [T_n, T_{n+1}]$, where $r(n) \stackrel{\text{def}}{=} \|\bar{x}(T_n)\| \vee 1, n \geq 0$. That is, we obtain $\hat{x}(\cdot)$ from $\bar{x}(\cdot)$ by observing the latter at times $\{T_n\}$ that are spaced approximately T apart. In case the observed value falls outside the unit ball of $\mathcal{R}^d$, it is reset to a value on the unit sphere of $\mathcal{R}^d$ by normalization. Not surprisingly, this prevents any possible blow-up of the trajectory, as reflected in the following lemma. For later use, we also define $\hat{x}(T_{n+1}^-) \stackrel{\text{def}}{=} \bar{x}(T_{n+1})/r(n)$. This is the same as $\hat{x}(T_{n+1})$ if there is no jump at T_{n+1}, and equal to $\lim_{t \uparrow T_{n+1}} \hat{x}(t)$ if there is a jump.

Lemma 4. $\sup_t E[\|\hat{x}(t)\|^2] < \infty$.

Proof. It suffices to show that

$$\sup_{m(n) \leq k < m(n+1)} E[\|\hat{x}(t(k))\|^2] < M$$

for some $M > 0$ independent of n.

Fix n. Then for $m(n) \leq k < m(n+1)$,

$$\hat{x}(t(k+1)) = \hat{x}(t(k)) + a(k)(h_{r(n)}(\hat{x}(t(k))) + \hat{M}_{k+1}),$$

where $\hat{M}_{k+1} \stackrel{\text{def}}{=} M_{k+1}/r(n)$. Since $r(n) \geq 1$, it follows from (A3) that $\hat{M}_{k+1}$ satisfies

$$E[\|\hat{M}_{k+1}\|^2 | \mathcal{F}_k] \leq K(1 + \|\hat{x}(t(k))\|^2). \tag{3.2.5}$$

Thus, $E[||\hat{M}_{k+1}||^2] \leq K(1 + E[||\hat{x}(t(k))||^2])$, which gives us the bound

$$E[||\hat{M}_{k+1}||^2]^{1/2} \leq \sqrt{K}(1 + E[||\hat{x}(t(k))||^2]^{1/2}).$$

(Note that for $a \geq 0$, $\sqrt{1 + a^2} \leq 1 + a$.) Using this and the bound $||h_c(x)|| \leq K_0(1 + ||x||)$ mentioned above, we have

$$E[||\hat{x}(t(k+1))||^2]^{\frac{1}{2}} \leq E[||\hat{x}(t(k))||^2]^{\frac{1}{2}}(1 + a(k)K_1) + a(k)K_2,$$

for suitable constants $K_1, K_2 > 0$. Keeping in mind that

$$\sum_{k=m(n)}^{m(n+1)-1} a(k) \leq T + 1, \quad ||\hat{x}(t(m(n)))|| \leq 1,$$

a straightforward recursion leads to

$$E[||\hat{x}(t(k+1))||^2]^{\frac{1}{2}} \leq e^{K_1(T+1)}(1 + K_2(T+1)),$$

where we also use the inequality $1 + x \leq e^x$. This is the desired bound. ∎

Lemma 5. *The sequence* $\hat{\zeta}_n \overset{def}{=} \sum_{k=0}^{n-1} a_k \hat{M}_{k+1}, n \geq 1$, *is a.s. convergent.*.

Proof. By the convergence theorem for square-integrable martingales (see Appendix C), it is enough to show that $\sum_k E[||a(k)\hat{M}_{k+1}||^2|\mathcal{F}_k] < \infty$ a.s. Thus it is enough to show that $E[\sum_k E[||a(k)\hat{M}_{k+1}||^2|\mathcal{F}_k]] < \infty$. Since, as in the proof of Lemma 4, $E[||\hat{M}_{k+1}||^2|\mathcal{F}_k] \leq K(1 + ||\hat{x}(t(k))||^2)$, we get

$$E[\sum_k E[||a(k)\hat{M}_{k+1}||^2|\mathcal{F}_k]] = \sum_k E[E[||a(k)\hat{M}_{k+1}||^2|\mathcal{F}_k]]$$

$$\leq \sum_k a(k)^2 K(1 + E[||\hat{x}(t(k))||^2]).$$

This is finite, by property (A2) and by Lemma 4. ∎

For $n \geq 0$, let $x^n(t), t \in [T_n, T_{n+1}]$, denote the trajectory of (3.2.3) with $c = r(n)$ and $x^n(T_n) = \hat{x}(T_n)$.

Lemma 6. $\lim_{n\to\infty} \sup_{t \in [T_n, T_{n+1}]} ||\hat{x}(t) - x^n(t)|| = 0$, *a.s.*

Proof. For simplicity, we assume $L > 1$, $a(n) < 1 \ \forall n$. Note that for $m(n) \leq k < m(n+1)$,

$$\hat{x}(t(k+1)) = \hat{x}(t(k)) + a(k)(h_{r(n)}(\hat{x}(t(k))) + \hat{M}_{k+1}).$$

This yields, for $0 < k \le m(n+1) - m(n)$,

$$\hat{x}(t(m(n) + k)) = \hat{x}(t(m(n))) + \sum_{i=0}^{k-1} a(m(n) + i)h_{r(n)}(\hat{x}(t(m(n) + i)))$$
$$+ (\hat{\zeta}_{m(n)+k} - \hat{\zeta}_{m(n)}).$$

By Lemma 5, there is a (random) bound B on $\sup_i \|\hat{\zeta}_i\|$. Also, as mentioned at the beginning of this section, we have

$$\|h_{r(n)}(\hat{x}(t(m(n) + i)))\| \le \|h(0)\| + L\|\hat{x}(t(m(n) + i))\|.$$

Furthermore, $\sum_{0 \le i < m(n+1) - m(n)} a(m(n) + i) \le (T+1)$. Therefore,

$$\|\hat{x}(t(m(n) + k))\|$$

$$\le \|\hat{x}(t(m(n)))\| + \sum_{i=0}^{k-1} a(m(n) + i)(\|h(0)\|$$
$$+ L\|\hat{x}(t(m(n) + i))\|) + 2B$$

$$\le L\sum_{i=0}^{k-1} a(m(n) + i)\|\hat{x}(t(m(n) + i))\| + \|h(0)\|(T+1)$$
$$+ 2B + 1,$$

where we use $\|\hat{x}(t(m(n)))\| \le 1$. We can now apply the discrete Gronwall inequality (see Appendix B) to obtain

$$\|\hat{x}(t(m(n) + k))\| \le (\|h(0)\|(T+1) + 2B + 1)e^{L(T+1)} \stackrel{\text{def}}{=} K^* \qquad (3.2.6)$$

for $0 < k \le m(n+1) - m(n)$. It follows that $\hat{x}$ remains bounded on $[T_n, T_{n+1}]$ by some $K^* > 0$ and this bound is independent of n. We can now mimic the argument of Lemma 1, Chapter 2, to show that

$$\lim_{n \to \infty} \sup_{t \in [T_n, T_{n+1}]} \|\hat{x}(t) - x^n(t)\| = 0, \quad \text{a.s.}$$

∎

This leads to our main result:

Theorem 7. *Under (A1)-(A3) and (A5),* $\sup_n \|x_n\| < \infty$ *a.s.*

Proof. Fix a sample point where the claims of Lemmas 5 and 6 hold. We will first show that $\sup_n \|\bar{x}(T_n)\| < \infty$. If this does not hold, there will exist a sequence $T_{n_1}, T_{n_2}, \ldots$ such that $\|\bar{x}(T_{n_k})\| \nearrow \infty$, i.e., $r_{n_k} \nearrow \infty$. We saw (Corollary 3) that there exists a scaling factor $c_0 > 0$ and a $T > 0$ such that for all initial conditions x on the unit sphere, $\|\phi_c(t, x)\| < \frac{1}{4}$ for $t \in [T, T+1]$

and $c > c_0$ (≥ 1 by assumption). If $r_n > c_0$, $\|\hat{x}(T_n)\| = \|x^n(T_n)\| = 1$, and $\|x^n(T_{n+1})\| < \frac{1}{4}$. But then by Lemma 6, $\|\hat{x}(T_{n+1}^-)\| < \frac{1}{2}$ if n is large. Thus, for $r_n > c_0$ and n sufficiently large,

$$\frac{\|\bar{x}(T_{n+1})\|}{\|\bar{x}(T_n)\|} = \frac{\|\hat{x}(T_{n+1}^-)\|}{\|\hat{x}(T_n)\|} < \frac{1}{2}.$$

We conclude that if $\|\bar{x}(T_n)\| > c_0$, $\bar{x}(T_k)$, $k \geq n$ falls back to the ball of radius c_0 at an exponential rate. Thus if $\|\bar{x}(T_n)\| > c_0$, $\|\bar{x}(T_{n-1})\|$ is either even greater than $\|\bar{x}(T_n)\|$ or is inside the ball of radius c_0. Then there must be an instance prior to n when $\bar{x}(\cdot)$ jumps from inside this ball to outside the ball of radius $0.9r_n$. Thus, corresponding to the sequence $r_{n_k} \nearrow \infty$, we will have a sequence of jumps of $\bar{x}(T_n)$ from inside the ball of radius c_0 to points increasingly far away from the origin. But, by a discrete Gronwall argument analogous to the one used in Lemma 6, it follows that there is a bound on the amount by which $\|\bar{x}(\cdot)\|$ can increase over an interval of length $T + 1$ when it is inside the ball of radius c_0 at the beginning of the interval. This leads to a contradiction. Thus $\tilde{C} \stackrel{\text{def}}{=} \sup_n \|\bar{x}(T_n)\| < \infty$. This implies that $\sup_n \|x_n\| \leq \tilde{C}K^* < \infty$ for K^* as in (3.2.6). ∎

Consider as an illustrative example the scalar case with $h(x) = -x + g(x)$ for some bounded Lipschitz g. Then $h_\infty(x) = -x$, indicating that the scaling limit $c \to \infty$ above basically picks the dominant term $-x$ of h that essentially controls the behaviour far away from the origin.

3.3 Another stability criterion

The second stability test we discuss is adapted from Abounady, Bertsekas and Borkar (2002). This applies to the case when stability for one initial condition implies stability for all initial conditions. Also, the associated o.d.e. (3.2.1) is assumed to converge to a bounded invariant set for all initial conditions. The idea then is to consider a related recursion 'with resets', i.e., a recursion which is reset to a bounded set whenever it exits from a larger prescribed bounded set containing the previous one. By a suitable choice of these sets (which explicitly depends on the dynamics), one then argues that there are at most finitely many resets. Hence the results of the preceding section apply thereafter. But this implies stability for *some* initial condition, hence for all initial conditions.

In this situation, it is more convenient to work with (1.0.4) of Chapter 1 rather than (1.0.3) there, i.e., with

$$x_{n+1} = x_n + a(n)f(x_n, \xi_{n+1}), \ n \geq 0, \tag{3.3.1}$$

where $\{\xi_n\}$ are i.i.d. random variables taking values in $\mathcal{R}^m$, say (though more general spaces can be admitted), and $f : \mathcal{R}^d \times \mathcal{R}^m \to \mathcal{R}^d$ satisfies

$$\|f(x, z) - f(y, z)\| \leq L\|x - y\|. \tag{3.3.2}$$

Thus $h(x) \stackrel{\text{def}}{=} E[f(x, \xi_1)]$ is Lipschitz with Lipschitz constant L and $M_{n+1} \stackrel{\text{def}}{=}$ $f(x_n, \xi_{n+1}) - h(x_n), n \geq 0$, is a martingale difference sequence satisfying (A3). This reduces (3.3.1) to the familiar form

$$x_{n+1} = x_n + a(n)[h(x_n) + M_{n+1}]. \tag{3.3.3}$$

We also assume that

$$\|f(x, y)\| \leq K(1 + \|x\|) \tag{3.3.4}$$

for a suitable $K > 0$. (The assumption simplifies matters, but could be replaced by other conditions that lead to similar conclusions.) Then

$$\|M_{n+1}\| \leq 2K(1 + \|x_n\|), \tag{3.3.5}$$

which implies (A3) of Chapter 2. The main assumption we shall be making is the following:

(*) Let $\{x'_n\}, \{x''_n\}$ be two sequences of random variables generated by the iteration (3.3.1) on a common probability space with the *same* 'driving noise' $\{\xi_n\}$, but different initial conditions. Then $\sup_n \|x'_n - x''_n\| < \infty$, a.s.

Typically in applications, $\sup_n \|x'_n - x''_n\|$ will be bounded by a function of $\|x'_0 - x''_0\|$. We further assume that (3.2.1) has an associated Liapunov function $V : \mathcal{R}^d \to \mathcal{R}$ that is continuously differentiable and satisfies:

(i) $\lim_{\|x\| \to \infty} V(x) = \infty$, and

(ii) for $\dot{V} \stackrel{\text{def}}{=} \langle \nabla V(x), h(x) \rangle$, $\dot{V} \leq 0$, and in addition, $\dot{V} < 0$ outside a bounded set $B \subset \mathcal{R}^d$.

It is worth noting here that we shall need only the existence and not the explicit form of V. The existence in turn is often ensured by smooth versions of the converse Liapunov theorem as in Wilson (1969).

Consider an initial condition x_0 (assumed to be deterministic for simplicity), and let G be a closed ball that contains in turn both $\bar{B}$ and x_0 in its interior. For $a \in \mathcal{R}$, let $C_a \stackrel{\text{def}}{=} \{x \in \mathcal{R}^d : V(x) \leq a\}$. V is bounded on G, so there exist b and c, with $b < c$ such that $G \subset C_b \subset C_c$. Choose $\delta < (c - b)/2$.

We define a modified iteration $\{x^*_n\}$ as follows: Let $x^*_0 = x_0$ and x^*_{n+1} be generated by (3.3.1) except when $x^*_n \notin C_c$. In the latter case we first replace x^*_n by its projection to the boundary ∂G of G and then continue. This is the 'reset' operation. Let $\tau_n, n \geq 1$, denote the successive reset times, with $+\infty$ being a possible value thereof. Let $\{t(n)\}$ be defined as before. Define

$\bar{x}(s)$ for $s \in [t(n), t(n+1)]$ as the linear interpolation of $\bar{x}(t(n)) = x_n^*$ and $\bar{x}(t(n+1)) = x_{n+1}^*$, using the post-reset value of x_n^* at $s = t(n)$ and the pre-reset value of x_{n+1}^* at $s = t(n+1)$ in case there is a reset at either time instant. Note that by (3.3.4), $\|\bar{x}(\cdot)\|$ remains bounded.

Now, we can choose a $\Delta > 0$ such that $\dot{V} \leq -\Delta$ on the compact set $\{x : b \leq V(x) \leq c\}$. Choose T such that $\Delta T > 2\delta$. Let $T_0 = 0$, and let $T_{n+1} = \min\{t(m) : t(m) \geq T_n + T\}$. For simplicity, suppose that $a(m) \leq 1$ for all m. Then $T_n + T \leq T_{n+1} \leq T_n + T + 1$.

For $t \in [T_n, T_{n+1}]$, define $x^n(t)$ to be the piecewise solution of $\dot{x} = h(x)$ such that $x^n(t)$ is set to $\bar{x}(t)$ at $t = T_n$ and also at $t = \tau_k$ for all $\tau_k \in [T_n, T_{n+1}]$. That is:

- it satisfies the o.d.e. on every subinterval $[t_1, t_2]$ where t_1 is either some T_n or a reset time in $[T_n, T_{n+1}]$ and t_2 is either its immediate successor from the set of reset times in (T_n, T_{n+1}) or T_{n+1}, whichever comes first, and
- at t_1 it equals $\bar{x}(t_1)$.

The following lemma is proved in a manner analogous to Lemma 1 of Chapter 2.

Lemma 8. *For $T > 0$,*

$$\lim_{n \to \infty} \sup_{t \in [T_n, T_{n+1}]} \|\bar{x}(t) - x^n(t)\| = 0, \ a.s.$$

Theorem 9. $\sup_n \|x_n\| < \infty$, *a.s.*

Proof. Let $x_k^* \in C_c$. Then by (3.3.4), there is a bounded set B_1 such that

$$x_{k+1}^{*,-} \stackrel{\text{def}}{=} x_k^* + a(k)[h(x_k^*) + M_{k+1}] \in B_1.$$

Let $\eta_1 > 0$. By Lemma 8, there is an N_1 such that for $n \geq N_1$, one has $\sup_{t \in [T_n, T_{n+1}]} \|\bar{x}(t) - x^n(t)\| < \eta_1$. Let D be the η_1-neighbourhood of B_1. Thus, for $n \geq N_1, t \in [T_n, T_{n+1}]$, both $\bar{x}(t)$ and $x^n(t)$ remain inside D. D has compact closure, and therefore V is uniformly continuous on D. Thus there is an $\eta_2 > 0$ such that for $x, y \in D$, $\|x - y\| < \eta_2$ implies $|V(x) - V(y)| < \delta$. Let $\eta = \min\{\eta_1, \eta_2\}$. Let $N \geq N_1$ be such that for $n \geq N$, $\sup_{t \in [T_n, T_{n+1}]} \|\bar{x}(t) - x^n(t)\| < \eta$. Then for $n \geq N, t \in [T_n, T_{n+1}]$, $x^n(t)$ remains inside D and furthermore, $|V(\bar{x}(t)) - V(x^n(t))| < \delta$.

Now fix $n \geq N$. Let $\tau_k \in [T_n, T_{n+1}]$ be a reset time. Let $t \in [\tau_k, T_{n+1}]$. Since $x^n(\tau_k)$ is on G, then $V(x^n(\tau_k)) \leq b$, and $V(x^n(t)), t \geq \tau_k$, decreases with t until T_{n+1} or until the next reset, if that occurs before T_{n+1}. At such a reset, however, $V(x^n(t))$ again has value at most b and the argument repeats. Thus in any case, $V(x^n(t)) \leq b$ on $[\tau_k, T_{n+1}]$. Thus $V(\bar{x}(t)) \leq b + \delta < c$ on $[\tau_k, T_{n+1}]$. Hence $\bar{x}(\cdot)$ does not exit C_c, and there can in fact be no further resets in $[\tau_k, T_{n+1}]$. Note also that $V(\bar{x}(T_{n+1})) \leq b + \delta$.

Now consider the interval $[T_{n+1}, T_{n+2}]$. Since $x^{n+1}(T_{n+1}) = \bar{x}(T_{n+1})$, we have $V(x^{n+1}(T_{n+1})) \leq b+\delta$. We argue as above to conclude that $V(x^{n+1}(t)) \leq b+\delta$ and $V(\bar{x}(t)) < b+2\delta < c$ on $[T_{n+1}, T_{n+2}]$. This implies that $\bar{x}(\cdot)$ does not leave C_c on $[T_{n+1}, T_{n+2}]$ and hence there are no resets on this interval. Further, if $x^{n+1}(\cdot)$ remained in the set $\{x : b \leq V(x) \leq c\}$, the rate of decrease of $V(x^{n+1}(t))$ would be at least Δ. Recall that $\Delta T > 2\delta$. Thus it must be that $V(x^{n+1}(T_{n+2})) \leq b$, which means that as before, $V(\bar{x}(T_{n+2})) \leq b+\delta$.

Repeating the argument for successive ns, it follows that beyond a certain point in time, there can be no further resets. Thus for n large enough, $\{x_n^*\}$ remains bounded and furthermore, $x_{n+1}^* = x_n^* + a(n)f(x_n^*, \xi_{n+1})$. But $x_{n+1} = x_n + a(n)f(x_n, \xi_{n+1})$, and the noise sequence is the same for both recursions. By the assumption (*), it follows that the sequence $\{x_n\}$ remains bounded, a.s.

If there is no reset after N, $\{x_n^*\}$ is anyway bounded and the same logic applies.   ■

This test of stability is useful in some reinforcement learning applications. Further stability criteria can be found, e.g., in Kushner and Yin (2003) and Tsitsiklis (1994).

4

Lock-in Probability

4.1 Estimating the lock-in probability

Recall the urn model of Chapter 1. When there are multiple isolated stable equilibria, it turns out that there can be a positive probability of convergence to one of these equilibria which is not, however, necessarily among the desired ones. This, we recall, was the explanation for several instances of adoption of one particular convention or technology as opposed to another. The idea is that after some initial randomness, the process becomes essentially 'locked into' the domain of attraction of a particular equilibrium, i.e., locked into a particular choice of technology or convention. With this picture in mind, we define the lock-in probability as the probability of convergence to an asymptotically stable attractor, given that the iterate is in a neighbourhood thereof. Our aim here will be to get a lower bound on this probability and explore some of its consequences. Our treatment follows the approach of Borkar (2002, 2003).

Our setting is as follows. We consider the stochastic approximation on $\mathcal{R}^d$:

$$x_{n+1} = x_n + a(n)[h(x_n) + M_{n+1}], \qquad (4.1.1)$$

under assumptions (A1), (A2) and (A3) of Chapter 2. We add to (A2) the requirement that $a(n) \leq ca(m) \ \forall \ n \geq m$ for some $c > 0$. We shall not a priori make assumption (A4) of Chapter 2, which says that the sequence $\{x_n\}$ generated by (4.1.1) is a.s. bounded. The a.s. boundedness of this sequence with high probability will be proved *as a consequence of our main result*. We have seen that recursion (4.1.1) can be considered to be a noisy discretization of the ordinary differential equation

$$\dot{x}(t) = h(x(t)). \qquad (4.1.2)$$

Let $G \subset \mathcal{R}^d$ be open, and let $V : G \to [0, \infty)$ be such that $\dot{V} \stackrel{\text{def}}{=} \nabla V \cdot h : G \to \mathcal{R}$

31

is non-positive. We shall assume that $H \stackrel{\text{def}}{=} \{x : V(x) = 0\}$ is equal to the set $\{x : \dot{V}(x) = 0\}$ and is a compact subset of G. Thus the function V is a Liapunov function. Then H is an asymptotically stable invariant set of the differential equation (4.1.2). Conversely, (local) asymptotic stability implies the existence of such a V by the converse Liapunov theorem – see, e.g., Krasovskii (1963). Let there be an open set B with compact closure such that $H \subset B \subset \bar{B} \subset G$. It follows from the LaSalle invariance principle (see Appendix B) that any internally chain transitive invariant subsets of $\bar{B}$ will be subsets of H.

In this setting we shall derive an estimate for the probability that the sequence $\{x_n\}$ is convergent to H, conditioned on the event that $x_{n_0} \in B$ for some n_0 sufficiently large. In the next section, we shall also derive a *sample complexity estimate* for the probability of the sequence being within a certain neighbourhood of H after a certain length of time, again conditioned on the event that $x_{n_0} \in B$.

Let $H^a \stackrel{\text{def}}{=} \{x : V(x) < a\}$ have compact closure $\bar{H}^a = \{x : V(x) \leq a\}$. For $A \subset \mathcal{R}^d, \delta > 0$, let $N_\delta(A)$ denote the δ-neighbourhood of A, i.e., $N_\delta(A) \stackrel{\text{def}}{=} \{x : \inf_{y \in A} \|x - y\| < \delta\}$. Fix some $0 < \epsilon_1 < \epsilon$ and $\delta > 0$ such that $N_\delta(H^{\epsilon_1}) \subset H^\epsilon \subset N_\delta(H^\epsilon) \subset B$.

As was argued in the first extension of Theorem 2, Chapter 2, in section 2.2, if the sequence $\{x_n\}$ generated by recursion (4.1.1) remains a.s. bounded on a prescribed set of sample points, then it converges almost surely on this set to a (possibly sample path dependent) compact internally chain transitive invariant set of the o.d.e. (4.1.2). Therefore, if we can show that with high probability $\{x_n\}$ remains inside the compact set $\bar{B}$, then it follows that $\{x_n\}$ converges to H with high probability. We shall in fact show that $\bar{x}(\cdot)$, the piecewise linear and continuous curve obtained by linearly interpolating the points $\{x_n\}$ as in Chapter 2, lies inside $N_\delta(H^\epsilon) \subset B$ with high probability from some time on. Let us define

$$T = \frac{[\max_{x \in \bar{B}} V(x)] - \epsilon_1}{\min_{x \in \bar{B} \setminus H^{\epsilon_1}} |\nabla V(x) \cdot h(x)|}.$$

Then T is an upper bound for the time required for a solution of the o.d.e. (4.1.2) to reach the set H^{ϵ_1}, starting from an initial condition in $\bar{B}$. Fix an $n_0 \geq 1$ 'sufficiently large'. (We shall be more specific later about how n_0 is to be chosen.) For $m \geq 1$, let $n_m = \min\{n : t(n) \geq t(n_{m-1}) + T\}$. Define a sequence of times $T_0, T_1, \ldots$ by $T_m = t(n_m)$. For $m \geq 0$, let I_m be the interval $[T_m, T_{m+1}]$, and let

$$\rho_m \stackrel{\text{def}}{=} \sup_{t \in I_m} \|\bar{x}(t) - x^{T_m}(t)\|,$$

where $x^{T_m}(\cdot)$ is, as in Chapter 2, the solution of the o.d.e. (4.1.2) on I_n with initial condition $x^{T_m}(T_m) = \bar{x}(T_m)$. We shall assume that $a(n) \leq 1\ \forall n$, which implies that the length of I_m is between T and $T + 1$.

Let us assume for the moment that $x_{n_0} \in B$, and that $\rho_m < \delta$ for all $m \geq 0$. Because of the way we defined T, it follows that $x^{T_0}(T_1) \in H^{\epsilon_1}$. Since $\rho_0 < \delta$ and $N_\delta(H^{\epsilon_1}) \subset H^\epsilon$, $\bar{x}(T_1) \in H^\epsilon$. Since H^ϵ is a positively invariant subset of $\bar{B}$, it follows that $x^{T_1}(\cdot)$ lies in H^ϵ on I_1, and that $x^{T_1}(T_2) \in H^{\epsilon_1}$. Hence $\bar{x}(T_2) \in H^\epsilon$. Continuing in this way it follows that for all $m \geq 1$, x^{T_m} lies inside H^ϵ on I_m. Now using the fact that $\rho_m < \delta$ for all $m \geq 0$, it follows that $\bar{x}(t)$ is in $N_\delta(H^\epsilon) \subset B$ for all $t \geq T_1$. As mentioned above, it now follows that $\bar{x}(t) \to H$ as $t \to \infty$. Therefore we have:

Lemma 1.

$$P(\bar{x}(t) \to H | x_{n_0} \in B) \geq P(\rho_m < \delta\ \forall m \geq 0 | x_{n_0} \in B). \qquad (4.1.3)$$

We let $\mathcal{B}_m$ denote the event that $x_{n_0} \in B$ and $\rho_k < \delta$ for $k = 0, 1, \ldots, m$. Recall that $\mathcal{F}_n \stackrel{\text{def}}{=} \sigma(x_0, M_1, \ldots, M_n), n \geq 1$. We then have that $\mathcal{B}_m \in \mathcal{F}_{n_{m+1}}$. Note that

$$P(\rho_m < \delta\ \forall m \geq 0 | x_{n_0} \in B) = 1 - P(\rho_m \geq \delta \text{ for some } m \geq 0 | x_{n_0} \in B)$$

We have the following disjoint union:

$$\{\rho_m \geq \delta \text{ for some } m \geq 0\} =$$
$$\{\rho_0 \geq \delta\} \cup \{\rho_1 \geq \delta;\ \rho_0 < \delta\} \cup \{\rho_2 \geq \delta;\ \rho_0, \rho_1 < \delta\} \cup \cdots$$

Therefore,

$$\begin{aligned}
&P(\rho_m \geq \delta \text{ for some } m \geq 0 | x_{n_0} \in B) \\
&= P(\rho_0 \geq \delta | x_{n_0} \in B) \\
&\quad + P(\rho_1 \geq \delta;\ \rho_0 < \delta | x_{n_0} \in B) \\
&\quad + P(\rho_2 \geq \delta;\ \rho_0, \rho_1 < \delta | x_{n_0} \in B) + \cdots \\
&= P(\rho_0 \geq \delta | x_{n_0} \in B) \\
&\quad + P(\rho_1 \geq \delta | \rho_0 < \delta, x_{n_0} \in B) P(\rho_0 < \delta | x_{n_0} \in B) \\
&\quad + P(\rho_2 \geq \delta | \rho_0, \rho_1 < \delta, x_{n_0} \in B) P(\rho_0, \rho_1 < \delta | x_{n_0} \in B) + \cdots \\
&\leq P(\rho_0 \geq \delta | x_{n_0} \in B) + P(\rho_1 \geq \delta | \mathcal{B}_0) + P(\rho_2 \geq \delta | \mathcal{B}_1) + \cdots
\end{aligned}$$

Thus, with $\mathcal{B}_{-1} \stackrel{\text{def}}{=} \{x_0 \in B\}$, we have:

Lemma 2.

$$P(\rho_m < \delta\ \forall m \geq 0 | x_{n_0} \in B) \geq 1 - \sum_{m=0}^{\infty} P(\rho_m \geq \delta | \mathcal{B}_{m-1}).$$

The bound derived in Lemma 2 involves the term $P(\rho_m \geq \delta | \mathcal{B}_{m-1})$. We shall now derive an upper bound on this term. Recall that $\mathcal{B}_{m-1}$ denotes the event that $x_{n_0} \in B$ and $\rho_k < \delta$ for $k = 0, 1, \ldots, m-1$. This implies that $\bar{x}(T_m) \in B$. Let C be a bound on $\|h(\Phi_t(x))\|$, where Φ_t is the time-t flow map for the o.d.e. (4.1.2), $0 \leq t \leq T+1$, and $x \in \bar{B}$. By the arguments of Lemma 1 of Chapter 2, it follows that

$$\rho_m \leq Ca(n_m) + K_T(CLb(n_m) + \max_{n_m \leq j \leq n_{m+1}} \|\zeta_j - \zeta_{n_m}\|),$$

where K_T is a constant that depends only on T, L is the Lipschitz constant for h, $b(n) \overset{\text{def}}{=} \sum_{k \geq n} a(k)^2$, and $\zeta_k = \sum_{i=0}^{k-1} a(i)M_{i+1}$. Since $a(n_m) \leq ca(n_0)$ and $b(n_m) \leq c^2 b(n_0)$, it follows that

$$\rho_m \leq (Ca(n_0) + cK_T CLb(n_0))c + K_T \max_{n_m \leq j \leq n_{m+1}} \|\zeta_j - \zeta_{n_m}\|.$$

This implies that if n_0 is chosen so that $(Ca(n_0) + cK_T CLb(n_0))c < \delta/2$, then $\rho_m > \delta$ implies that

$$\max_{n_m \leq j \leq n_{m+1}} \|\zeta_j - \zeta_{n_m}\| > \frac{\delta}{2K_T}.$$

We state this as a lemma:

Lemma 3. *If n_0 is chosen so that*

$$(Ca(n_0) + cK_T CLb(n_0))c < \delta/2, \tag{4.1.4}$$

then

$$P(\rho_m \geq \delta | \mathcal{B}_{m-1}) \leq P(\max_{n_m \leq j \leq n_{m+1}} \|\zeta_j - \zeta_{n_m}\| > \frac{\delta}{2K_T} | \mathcal{B}_{m-1})$$

We shall now find a bound for the expression displayed on the right-hand side of the above inequality. We shall give two methods for bounding this quantity. The first one uses Burkholder's inequality, and the second uses a concentration inequality for martingales. As we shall see, the second method gives a better bound, but under a stronger assumption.

Burkholder's inequality (see Appendix C) implies that if $(X_n, \mathcal{F}_n)$, $n \geq 1$ is a (real-valued) zero mean martingale, and if $\bar{M}_n \overset{\text{def}}{=} X_n - X_{n-1}$ is the corresponding martingale difference sequence, then there is a constant C_1 such that

$$E[(\max_{0 \leq j \leq n} \|X_j\|)^2] \leq C_1^2 E[\sum_{i=1}^{n} \bar{M}_i^2].$$

We shall use the conditional version of this inequality: If $\mathcal{G} \subset \mathcal{F}_0$ is a sub-σ-field, then

$$E[(\max_{0 \leq j \leq n} \|X_j\|)^2 | \mathcal{G}] \leq C_1^2 E[\sum_{i=1}^{n} \bar{M}_i^2 | \mathcal{G}]$$

almost surely. In our context, $\{\zeta_j - \zeta_{n_m}, j \geq n_m\}$ is an $\mathcal{R}^d$-valued martingale with respect to the filtration $\{\mathcal{F}_j\}$. We shall apply Burkholder's inequality to each component.

We shall first prove some useful lemmas. Let K be the constant in assumption (A3) of Chapter 2. For an $\mathcal{R}^d$-valued random variable x, let $\|x\|^*$ denote $E[\|x\|^2|\mathcal{B}_{m-1}]^{1/2}(\omega)$, where $\omega \in \mathcal{B}_{m-1}$. Note that $\|\cdot\|^*$ satisfies the properties of a norm 'almost surely'.

Lemma 4. *For $n_m < j \leq n_{m+1}$ and a.s. $\omega \in \mathcal{B}_{m-1}$,*

$$(\|M_j\|^*)^2 \leq K(1 + (\|x_{j-1}\|^*)^2).$$

Proof. Note that $\mathcal{B}_{m-1} \in \mathcal{F}_{n_m} \subset \mathcal{F}_{j-1}$ for $j > n_m$. Then a.s.,

$$\begin{aligned}
(\|M_j\|^*)^2 &= E[\|M_j\|^2|\mathcal{B}_{m-1}](\omega) \\
&= E[E[\|M_j\|^2|\mathcal{F}_{j-1}]|\mathcal{B}_{m-1}](\omega) \\
&\leq E[K(1 + \|x_{j-1}\|^2)|\mathcal{B}_{m-1}](\omega) \\
&\qquad \text{(by Assumption (A3) of Chapter 2)} \\
&= K(1 + E[\|x_{j-1}\|^2|\mathcal{B}_{m-1}](\omega)) \\
&= K(1 + (\|x_{j-1}\|^*)^2).
\end{aligned}$$

∎

Lemma 5. *There is a constant $\bar{K}_T$ such that for $n_m \leq j \leq n_{m+1}$,*

$$\|x_j\|^* \leq \bar{K}_T \text{ a.s.}$$

Proof. Consider the recursion

$$x_{j+1} = x_j + a(j)[h(x_j) + M_{j+1}], \; j \geq n_m.$$

As we saw in Chapter 2, the Lipschitz property of h implies a linear growth condition on h, i.e., $\|h(x)\| \leq K'(1 + \|x\|)$. Taking the Euclidean norm on both sides of the above equation and using the triangle inequality and this linear growth condition leads to

$$\|x_{j+1}\| \leq \|x_j\|(1 + a(j)K') + a(j)K' + a(j)\|M_{j+1}\|.$$

Therefore a.s. for $j \geq n_m$,

$$\begin{aligned}
\|x_{j+1}\|^* &\leq \|x_j\|^*(1 + a(j)K') + a(j)K' + a(j)\|M_{j+1}\|^* \\
&\leq \|x_j\|^*(1 + a(j)K') + a(j)K' + a(j)\sqrt{K}(1 + (\|x_j\|^*)^2)^{1/2} \\
&\qquad \text{(by the previous lemma)} \\
&\leq \|x_j\|^*(1 + a(j)K') + a(j)K' + a(j)\sqrt{K}(1 + \|x_j\|^*) \\
&= \|x_j\|^*(1 + a(j)K_1) + a(j)K_1,
\end{aligned}$$

$$\text{where } K_1 = K' + \sqrt{K}.$$

Applying this inequality repeatedly and using the fact that $1+aK_1 \leq e^{aK_1}$ and the fact that if $j < n_{m+1}$, $a(n_m)+a(n_m+1)+\cdots+a(j) \leq t(n_{m+1})-t(n_m) \leq T+1$, we get

$$\|x_{j+1}\|^* \leq \|x_{n_m}\|^* e^{K_1(T+1)} + K_1(T+1)e^{K_1(T+1)} \text{ for } n_m \leq j < n_{m+1}.$$

For $\omega \in \mathcal{B}_{m-1}$, $x_{n_m}(\omega) \in B$. Since $\bar{B}$ is compact, there is a constant K_2 such that $\|x_{n_m}\| < K_2$. Thus $\|x_{n_m}\|^* \leq K_2$. This implies that for $n_m - 1 \leq j < n_{m+1}$, $\|x_{j+1}\|^* \leq \bar{K}_T \stackrel{\text{def}}{=} e^{K_1(T+1)}[K_2 + K_1(T+1)]$. In other words, for $n_m \leq j \leq n_{m+1}$, $\|x_j\|^* \leq \bar{K}_T$. ∎

Lemma 6. *For* $n_m < j \leq n_{m+1}$,

$$(\|M_j\|^*)^2 \leq K(1 + \bar{K}_T^2).$$

Proof. This follows by combining the results of Lemmas 4 and 5. ∎

For a.s. $\omega \in \mathcal{B}_{m-1}$, we have

$$P(\max_{n_m \leq j \leq n_{m+1}} \|\zeta_j - \zeta_{n_m}\| > \frac{\delta}{2K_T}|\mathcal{B}_{m-1})$$

$$= P((\max_{n_m \leq j \leq n_{m+1}} \|\zeta_j - \zeta_{n_m}\|)^2 > \frac{\delta^2}{4K_T^2}|\mathcal{B}_{m-1})$$

$$\leq E[(\max_{n_m \leq j \leq n_{m+1}} \|\zeta_j - \zeta_{n_m}\|)^2|\mathcal{B}_{m-1}] \cdot \frac{4K_T^2}{\delta^2},$$

where the inequality follows from the conditional Chebyshev inequality. We shall now use Burkholder's inequality to get the desired bound. Let ζ_n^i denote the ith component of ζ_n. For a.s. $\omega \in \mathcal{B}_{m-1}$,

$$E[(\max_{n_m \leq j \leq n_{m+1}} \|\zeta_j - \zeta_{n_m}\|)^2 | \mathcal{B}_{m-1}]$$

$$= E[\max_{n_m \leq j \leq n_{m+1}} \sum_{i=1}^{d} (\zeta_j^i - \zeta_{n_m}^i)^2 | \mathcal{B}_{m-1}]$$

$$\leq \sum_{i=1}^{d} E[(\max_{n_m \leq j \leq n_{m+1}} |\zeta_j^i - \zeta_{n_m}^i|)^2 | \mathcal{B}_{m-1}]$$

$$\leq \sum_{i=1}^{d} C_1^2 E[\sum_{j=n_m+1}^{n_{m+1}} a(j-1)^2 (M_j^i)^2 | \mathcal{B}_{m-1}]$$

$$= C_1^2 E[\sum_{j=n_m+1}^{n_{m+1}} a(j-1)^2 \|M_j\|^2 | \mathcal{B}_{m-1}]$$

$$= C_1^2 \sum_{j=n_m+1}^{n_{m+1}} a(j-1)^2 (\|M_j\|^*)^2$$

$$\leq C_1^2 [a_{n_m}^2 + \cdots + a_{n_{m+1}-1}^2] K(1 + \bar{K}_T^2)$$

$$= C_1^2 (b(n_m) - b(n_{m+1})) K(1 + \bar{K}_T^2).$$

Combining the foregoing, we obtain the following lemma:

Lemma 7. *For* $\tilde{K} \stackrel{def}{=} 4C_1^2 K(1 + \bar{K}_T^2) K_T^2 > 0,$

$$P(\max_{n_m \leq j \leq n_{m+1}} \|\zeta_j - \zeta_{n_m}\| > \frac{\delta}{2K_T} | \mathcal{B}_{m-1}) \leq \frac{\tilde{K}}{\delta^2} (b(n_m) - b(n_{m+1})).$$

Thus we have:

Theorem 8. *For some constant* $\tilde{K} > 0,$

$$P(\bar{x}(t) \to H | x_{n_0} \in B) \geq 1 - \frac{\tilde{K}}{\delta^2} b(n_0).$$

Proof. Note that

$$P(\rho_m < \delta \; \forall m \geq 0 | x_{n_0} \in B)$$

$$\geq 1 - \sum_{m=0}^{\infty} P(\rho_m \geq \delta | \mathcal{B}_{m-1}) \text{ by Lemma 2}$$

$$\geq 1 - \sum_{m=0}^{\infty} P(\max_{n_m \leq j \leq n_{m+1}} ||\zeta_j - \zeta_{n_m}|| > \frac{\delta}{2K_T} | \mathcal{B}_{m-1}) \text{ by Lemma 3}$$

$$\geq 1 - \sum_{m=0}^{\infty} \frac{\tilde{K}}{\delta^2} (b(n_m) - b(n_{m+1})) \text{ by Lemma 7}$$

$$= 1 - \frac{\tilde{K}}{\delta^2} b(n_0).$$

The claim now follows from Lemma 1.   ∎

A key assumption for the bound derived in Theorem 8 was assumption (A3) of Chapter 2, by which $E[||M_j||^2 | \mathcal{F}_{j-1}] \leq K(1 + ||x_{j-1}||^2)$, for $j \geq 1$. Now we shall make the more restrictive assumption that $||M_j|| \leq K_0(1 + ||x_{j-1}||)$. This condition holds, e.g., in the reinforcement learning applications discussed in Chapter 10. Under this assumption, it will be possible for us to derive a sharper bound.

We first prove a lemma about the boundedness of the stochastic approximation iterates:

Lemma 9. *There is a constant K_3 such that for $t \in I_m$, $||\bar{x}(t)|| \leq K_3(1 + ||\bar{x}(T_m)||)$.*

Proof. The stochastic approximation recursion is

$$x_{n+1} = x_n + a(n)[h(x_n) + M_{n+1}].$$

Therefore, assuming that $||M_j|| \leq K_0(1 + ||x_{j-1}||)$ and using the linear growth property of h,

$$||x_{n+1}|| \leq ||x_n|| + a(n)K(1 + ||x_n||) + a(n)K_0(1 + ||x_n||)$$

$$= ||x_n||(1 + a(n)K_4) + a(n)K_4, \text{ where } K_4 \overset{\text{def}}{=} K + K_0.$$

Arguing as in Lemma 5, we conclude that for $n_m \leq j < n_{m+1}$,

$$||x_{j+1}|| \leq [||x_{n_m}|| + K_4(T + 1)]e^{K_4(T+1)}.$$

Thus for $n_m \leq j \leq n_{m+1}$, $||x_j|| \leq e^{K_4(T+1)}[||x_{n_m}|| + K_4(T + 1)] \leq K_3(1 + ||x_{n_m}||)$ for some constant K_3. The lemma follows.   ∎

Note that for $n_m \leq k < n_{m+1}$,

$$
\begin{aligned}
\|\zeta_{k+1} - \zeta_k\| &= \|a(k)M_{k+1}\| \\
&\leq a(k)K_0(1 + \|x_k\|) \\
&\leq a(k)K_0(1 + K_3(1 + \|x_{n_m}\|)) \\
&\leq a(k)K_0[1 + K_3(K_2 + 1)], \\
&\qquad\qquad \text{where } K_2 \stackrel{\text{def}}{=} \max_{x \in \bar{B}} \|x\| \\
&= a(k)\bar{K}, \qquad\qquad\qquad\qquad\qquad\qquad (4.1.5) \\
&\qquad\qquad \text{where } \bar{K} \stackrel{\text{def}}{=} K_0[1 + K_3(K_2 + 1)].
\end{aligned}
$$

Thus one may use the following 'concentration inequality for martingales' (cf. Appendix C): Consider the filtration $\mathcal{F}_1 \subset \mathcal{F}_2 \subset \cdots \subset \mathcal{F}$. Let $S_1, \ldots, S_n$ be a (scalar) martingale with respect to this filtration, with $Y_1 = S_1$, $Y_k = S_k - S_{k-1}$ $(k \geq 2)$ the corresponding martingale difference sequence. Let $c_k \leq Y_k \leq b_k$. Then

$$
P(\max_{1 \leq k \leq n} |S_k| \geq t) \leq 2e^{\frac{-2t^2}{\sum_{k \leq n}(b_k - c_k)^2}}.
$$

If $\mathcal{B} \in \mathcal{F}_1$, we can state a conditional version of this inequality as follows:

$$
P(\max_{1 \leq k \leq n} |S_k| \geq t | \mathcal{B}) \leq 2e^{\frac{-2t^2}{\sum_{k \leq n}(b_k - c_k)^2}}.
$$

Let $\|\cdot\|_\infty$ denote the max-norm on $\mathcal{R}^d$, i.e., $\|x\|_\infty \stackrel{\text{def}}{=} \max_i |x_i|$. Note that for $v \in \mathcal{R}^d$, $\|v\|_\infty \leq \|v\| \leq \sqrt{d}\|v\|_\infty$. Thus $\|v\| \geq c$ implies that $\|v\|_\infty \geq c/\sqrt{d}$.

Since $\{\zeta_j - \zeta_{n_m}\}_{n_m \leq j \leq n_{m+1}}$ is a martingale with respect to $\{\mathcal{F}_j\}_{n_m \leq j \leq n_{m+1}}$ and $\mathcal{B}_{m-1} \in \mathcal{F}_{n_m}$, we have, by the inequality (4.1.5),

$$P(\max_{n_m \leq j \leq n_{m+1}} \|\zeta_j - \zeta_{n_m}\| > \frac{\delta}{2K_T}|\mathcal{B}_{m-1})$$

$$\leq P(\max_{n_m \leq j \leq n_{m+1}} \|\zeta_j - \zeta_{n_m}\|_\infty > \frac{\delta}{2K_T\sqrt{d}}|\mathcal{B}_{m-1})$$

$$= P(\max_{n_m \leq j \leq n_{m+1}} \max_{1 \leq i \leq d} |\zeta_j^i - \zeta_{n_m}^i| > \frac{\delta}{2K_T\sqrt{d}}|\mathcal{B}_{m-1})$$

$$= P(\max_{1 \leq i \leq d} \max_{n_m \leq j \leq n_{m+1}} |\zeta_j^i - \zeta_{n_m}^i| > \frac{\delta}{2K_T\sqrt{d}}|\mathcal{B}_{m-1})$$

$$\leq \sum_{i=1}^d P(\max_{n_m \leq j \leq n_{m+1}} |\zeta_j^i - \zeta_{n_m}^i| > \frac{\delta}{2K_T\sqrt{d}}|\mathcal{B}_{m-1})$$

$$\leq \sum_{i=1}^d 2\exp\left\{-2\frac{\delta^2/(4K_T^2 d)}{4[a(n_m)^2 + \cdots + a(n_{m+1}-1)^2]\bar{K}^2}\right\}$$

$$\leq 2d\exp\left\{-\frac{\delta^2}{8K_T^2\bar{K}^2 d[b(n_m) - b(n_{m+1})]}\right\}$$

$$= 2de^{-\hat{C}\delta^2/(d[b(n_m)-b(n_{m+1})])}, \text{ where } \hat{C} = 1/(8K_T^2\bar{K}^2).$$

This gives us:

Lemma 10. *There is a constant $\hat{C} > 0$ such that*

$$P(\max_{n_m \leq j \leq n_{m+1}} \|\zeta_j - \zeta_{n_m}\| > \frac{\delta}{2K_T}|\mathcal{B}_{m-1}) \leq 2de^{-\hat{C}\delta^2/(d[b(n_m)-b(n_{m+1})])}.$$

For n_0 sufficiently large that (4.1.4) holds and

$$b(n_0) < \hat{C}\delta^2/d, \tag{4.1.6}$$

the following bound holds:

Theorem 11.

$$P(\rho_m < \delta \,\forall m \geq 0 | x_{n_0} \in B) \geq 1 - 2de^{-\frac{\hat{C}\delta^2}{db(n_0)}} = 1 - o(b(n_0)).$$

Proof. Note that Lemmas 2 and 3 continue to apply. Then

$$P(\rho_m < \delta \,\forall m \geq 0 | x_{n_0} \in B)$$

$$\geq 1 - \sum_{m=0}^\infty P(\rho_m \geq \delta | \mathcal{B}_{m-1}) \text{ by Lemma 2}$$

$$\geq 1 - \sum_{m=0}^\infty P(\max_{n_m \leq j \leq n_{m+1}} \|\zeta_j - \zeta_{n_m}\| > \frac{\delta}{2K_T}|\mathcal{B}_{m-1}) \text{ by Lemma 3}$$

$$\geq 1 - \sum_{m=0}^\infty 2de^{-\hat{C}\delta^2/(d[b(n_m)-b(n_{m+1})])} \text{ by Lemma 10}$$

Note that for $C' > 0$, $e^{-C'/x}/x \to 0$ as $x \to 0$ and increases with x for $0 < x < C'$. Therefore for sufficiently large n_0,

$$\frac{e^{-\hat{C}\delta^2/(d[b(n_m)-b(n_{m+1})])}}{[b(n_m) - b(n_{m+1})]} \leq \frac{e^{-\hat{C}\delta^2/(db(n_m))}}{b(n_m)} \leq \frac{e^{-\hat{C}\delta^2/(db(n_0))}}{b(n_0)}.$$

Hence

$$e^{-\hat{C}\delta^2/(d[b(n_m)-b(n_{m+1})])} = [b(n_m) - b(n_{m+1})] \cdot \frac{e^{-\hat{C}\delta^2/(d[b(n_m)-b(n_{m+1})])}}{[b(n_m) - b(n_{m+1})]}$$

$$\leq [b(n_m) - b(n_{m+1})] \cdot \frac{e^{-\hat{C}\delta^2/(db(n_0))}}{b(n_0)}.$$

So,

$$\sum_{m=0}^{\infty} e^{-\hat{C}\delta^2/(d[b(n_m)-b(n_{m+1})])} \leq \sum_{m=0}^{\infty} [b(n_m) - b(n_{m+1})] \cdot \frac{e^{-\hat{C}\delta^2/(db(n_0))}}{b(n_0)}$$

$$= \frac{e^{-\hat{C}\delta^2/(db(n_0))}}{b(n_0)} \sum_{m=0}^{\infty} [b(n_m) - b(n_{m+1})]$$

$$= e^{-\hat{C}\delta^2/(db(n_0))}$$

The claim follows.   ∎

Lemma 1 coupled with Theorems 8 and 11 now enables us to derive bounds on the lock-in probability. We state these bounds in the following corollary:

Corollary 12. *In the setting described at the beginning of this section, if n_0 is chosen so that (4.1.4) holds, then there is a constant $\tilde{K}$ such that*

$$P(\bar{x}(t) \to H | x_{n_0} \in B) \geq 1 - \frac{\tilde{K}}{\delta^2} b(n_0).$$

If we make the additional assumption that $\|M_j\| \leq K_0(1 + \|x_{j-1}\|)$ for $j \geq 1$ and (4.1.6) holds, then the following tighter bound for the lock-in probability holds:

$$P(\bar{x}(t) \to H | x_{n_0} \in B) \geq 1 - 2de^{-\frac{\hat{C}\delta^2}{db(n_0)}}$$

$$= 1 - o(b(n_0)).$$

In conclusion, we observe that the stronger assumption on the martingale difference sequence $\{M_n\}$ was made necessary by the fact that we use McDiarmid's concentration inequality for martingales, which requires the associated martingale difference sequence to be bounded by deterministic bounds. More recent work on concentration inequalities for martingales may be used to relax this condition. See, e.g., Li (2003).

4.2 Sample complexity

We continue in the setting described at the beginning of the previous section. Our goal is to derive a *sample complexity estimate*, by which we mean an estimate of the probability that $\bar{x}(t)$ is within a certain neighbourhood of H after the lapse of a certain amount of time, conditioned on the event that $x_{n_0} \in B$ for some fixed n_0 sufficiently large.

We begin by fixing some $\epsilon > 0$ such that $\bar{H}^\epsilon \ (= \{x : V(x) \le \epsilon\}) \subset B$. Fix some $T > 0$, and let

$$\Delta \stackrel{\text{def}}{=} \min_{x \in \bar{B} \setminus H^\epsilon} [V(x) - V(\Phi_T(x))],$$

where Φ_T is the time-T flow map of the o.d.e. (4.1.2) (see Appendix B). Note that $\Delta > 0$. We remark that the arguments that follow do not require V to be differentiable, as was assumed earlier. It is enough to assume that V is continuous and that $V(x(t))$ monotonically decreases with t along any trajectory of (4.1.2) in $B \setminus H$. One such situation with non-differentiable V will be encountered in Chapter 10, in the context of reinforcement learning.

We fix an $n_0 \ge 1$ sufficiently large. We shall specify later how large n_0 needs to be. Let n_m, T_m, I_m, ρ_m and $\mathcal{B}_m$ be defined as in the previous section. Fix a $\delta > 0$ such that $N_\delta(H^\epsilon) \subset B$ and such that for all $x, y \in \bar{B}$ with $\|x - y\| < \delta$, $\|V(x) - V(y)\| < \Delta/2$.

Let us assume that $x_{n_0} \in B$, and that $\rho_m < \delta$ for all $m \ge 0$. If $x_{n_0} \in B \setminus H^\epsilon$, we have that $V(x^{T_0}(T_1)) \le V(\bar{x}(T_0)) - \Delta$. Since $\|x^{T_0}(T_1) - \bar{x}(T_1)\| < \delta$, it follows that $V(\bar{x}(T_1)) \le V(\bar{x}(T_0)) - \Delta/2$. If $\bar{x}(T_1) \in B \setminus H^\epsilon$, the same argument can be repeated to give $V(\bar{x}(T_2)) \le V(\bar{x}(T_1)) - \Delta/2$. Since $V(\bar{x}(T_m))$ cannot decrease at this rate indefinitely, it follows that $\bar{x}(T_{m_0}) \in H^\epsilon$ for some m_0. In fact, if

$$\tau \stackrel{\text{def}}{=} \frac{[\max_{x \in \bar{B}} V(x)] - \epsilon}{\Delta/2} \cdot (T+1),$$

then $T_{m_0} \le T_0 + \tau$.

Thus $x^{T_{m_0}}(t) \in H^\epsilon$ on $I_{m_0} = [T_{m_0}, T_{m_0+1}]$. Therefore $\bar{x}(T_{m_0+1}) \in H^{\epsilon+\Delta/2}$. This gives rise to two possibilities: either $\bar{x}(T_{m_0+1}) \in H^\epsilon$ or $\bar{x}(T_{m_0+1}) \in H^{\epsilon+\Delta/2} \setminus H^\epsilon$. In the former case, $x^{T_{m_0+1}}(t) \in H^\epsilon$ on I_{m_0+1}, and $\bar{x}(T_{m_0+2}) \in H^{\epsilon+\Delta/2}$. In the latter case, $x^{T_{m_0+1}}(t) \in H^{\epsilon+\Delta/2}$ on I_{m_0+1}, $x^{T_{m_0+1}}(T_{m_0+2}) \in H^{\epsilon-\Delta/2} \subset H^\epsilon$, and again $\bar{x}(T_{m_0+2}) \in H^{\epsilon+\Delta/2}$. In any case, $\bar{x}(T_{m_0+1}) \in H^{\epsilon+\Delta/2}$ implies that $x^{T_{m_0+1}}(t) \in H^{\epsilon+\Delta/2}$ on I_{m_0+1} and $\bar{x}(T_{m_0+2}) \in H^{\epsilon+\Delta/2}$. This argument can be repeated. We have thus shown that if $\bar{x}(T_{m_0}) \in H^\epsilon$, then $x^{T_{m_0+k}}(t) \in H^{\epsilon+\Delta/2}$ on I_{m_0+k} for all $k \ge 0$, which in turn implies that $\bar{x}(t) \in N_\delta(H^{\epsilon+\Delta/2})$ for all $t \ge T_{m_0}$. We thus conclude that if $x_{n_0} \in B$ and $\rho_m < \delta$ for all $m \ge 0$, then $\bar{x}(t) \in N_\delta(H^{\epsilon+\Delta/2})$ for all $t \ge T_{m_0}$, and thus for all $t \ge t(n_0) + \tau$. This gives us:

Lemma 13.

$$P(\bar{x}(t) \in N_\delta(H^{\epsilon+\Delta/2}) \; \forall t \geq t(n_0) + \tau | x_{n_0} \in B)$$
$$\geq P(\rho_m < \delta \; \forall m \geq 0 | x_{n_0} \in B)$$

Lemma 13 coupled with Theorems 8 and 11 now allows us to derive the following sample complexity estimate.

Corollary 14. *In the setting described at the beginning of section 4.1, if n_0 is chosen so that (4.1.4) holds, then there is a constant $\tilde{K}$ such that*

$$P(\bar{x}(t) \in N_\delta(H^{\epsilon+\Delta/2}) \; \forall t \geq t(n_0) + \tau | x_{n_0} \in B) \geq 1 - \frac{\tilde{K}}{\delta^2} b(n_0).$$

If we make the additional assumption that $\|M_j\| \leq K_0(1 + \|x_{j-1}\|)$ for $j \geq 1$ and (4.1.6) holds, then there is a constant C such that the following tighter bound holds:

$$
\begin{aligned}
P(\bar{x}(t) \in N_\delta(H^{\epsilon+\Delta/2}) \; \forall t \geq t(n_0) + \tau | x_{n_0} \in B) \; &\geq \; 1 - 2de^{-\frac{C\delta^2}{db(n_0)}} \\
&= \; 1 - o(b(n_0)).
\end{aligned}
$$

The corollary clearly gives a sample complexity type result in the sense described at the beginning of this section. There is, however, a subtle difference from the traditional sample complexity results: What we present here is not the number of samples needed to get within a prescribed accuracy with a prescribed probability starting from time zero, but starting from time n_0, and the bound depends crucially on the position at time n_0 via its dependence on the set B that we are able to choose.

As an example, consider the situation when $h(x) = g(x) - x$, where $g(\cdot)$ is a contraction, so that $\|g(x) - g(y)\| < \alpha\|x - y\|$ for some $\alpha \in (0,1)$. Let x^* be the unique fixed point of $g(\cdot)$, guaranteed by the contraction mapping theorem (see Appendix A). Straightforward calculation shows that $V(x) = \|x - x^*\|$ satisfies our requirements: Let

$$\dot{X}(x,t) = h(X(x,t)), \; X(x,0) = x.$$

We have

$$(X(x,t) - x^*) = (x - x^*) + \int_0^t (g(X(x,s)) - x^*)ds - \int_0^t (X(x,s) - x^*)ds,$$

leading to

$$(X(x,t) - x^*) = e^{-t}(x - x^*) + \int_0^t e^{-(t-s)}(g(X(x,s)) - x^*)ds.$$

Taking norms and using the contraction property,

$$\|X(x,t) - x^*\| \leq e^{-t}\|x - x^*\| + \int_0^t e^{-(t-s)}\|g(X(x,s)) - x^*\|ds$$

$$\leq e^{-t}\|x - x^*\| + \alpha \int_0^t e^{-(t-s)}\|X(x,s) - x^*\|ds.$$

That is,

$$e^t\|X(x,t) - x^*\| \leq \|x - x^*\| + \alpha \int_0^t e^s\|X(x,s) - x^*\|ds.$$

By the Gronwall inequality,

$$\|X(x,t) - x^*\| \leq e^{-(1-\alpha)t}\|x - x^*\|.$$

Let the iteration be at a point $\bar{x}$ at time n_0 large enough that (4.1.4) and (4.1.6) hold. Let $\|\bar{x} - x^*\| = b$ (say). For $\epsilon, T > 0$ as above, one may choose $\Delta = \epsilon(1 - e^{-(1-\alpha)T})$. We may take $\delta = \frac{\Delta}{2} \leq \frac{\epsilon}{2}$. One then needs

$$N_0 \stackrel{\text{def}}{=} \min \left\{ n : \sum_{i=n_0+1}^{n} a(i) \geq \frac{2(T+1)b}{\epsilon(1 - e^{-(1-\alpha)T})} \right\}$$

more iterates to get within 2ϵ of x^*, with a probability exceeding

$$1 - 2de^{-\frac{c\epsilon^2}{b(m(n_0))}} = 1 - o(b(n_0))$$

for a suitable constant c that depends on T, among other things. Note that $T > 0$ is a free parameter affecting both the expression for N_0 and that for the probability.

4.3 Avoidance of traps

As a second application of the estimates for the lock-in probability, we shall prove the *avoidance of traps* under suitable conditions. This term refers to the fact that under suitable additional hypotheses, the stochastic approximation iterations asymptotically avoid with probability one the attractors which are unstable in some direction. As one might guess, the additional hypotheses required concern the behaviour of h in the immediate neighbourhood of these attractors and a 'richness' condition on the noise. Intuitively, there should be an unstable direction at all length scales and the noise should be rich enough that it pushes the iterates in such a direction sufficiently often. This in turn ensures that they are eventually pushed away for good.

The importance of these results stems from the following considerations: We know from Chapter 2 that invariant sets of the o.d.e. are candidate limit sets for the algorithm. In many applications, the unstable invariant sets are precisely the spurious or undesirable limit sets one wants to avoid. The results of this section then give conditions when that avoidance will be achieved. More generally, these results allow us to narrow down the search for possible limit sets. As before, we work with the hypothesis (A4) of Chapter 2:

$$\sup_n \|x_n\| < \infty \text{ a.s.}$$

Consider a scenario where there exists an invariant set of (4.1.2) which is a disjoint union of N compact attractors $A_i, 1 \leq i \leq N$, with domains of attraction $G_i, 1 \leq i \leq N$, resp., such that $G = \bigcup_i G_i$ is open dense in $\mathcal{R}^d$. Let $W = G^c$. We shall impose further conditions on W as follows. Let D_α denote the truncated (open) cone

$$\{x = [x_1, \ldots, x_d] \in \mathcal{R}^d : 1 < x_1 < 2, |\sum_{i=2}^d x_i^2|^{\frac{1}{2}} < \alpha x_1\}$$

for some $\alpha > 0$. For a $d \times d$ orthogonal matrix O, $x \in \mathcal{R}^d$ and $a > 0$, let OD_α, $x + D_\alpha$ and aD_α denote resp. the rotation of D_α by O, translation of D_α by x, and scaling of D_α by a. Finally, for $\epsilon > 0$, let W_ϵ denote the ϵ-neighbourhood of W in $\mathcal{R}^d$. We shall be making some additional assumptions regarding (4.1.1) over and above those already in place. Our main additional assumption will be:

(A5) There exists an $\alpha > 0$ such that for any $a > 0$ and $x \in \mathcal{R}^d$, there exists an orthogonal matrix $O_{x,a}$ such that $x + aO_{x,a}D_\alpha \subset W_a^c$.

What this means is that for any $a > 0$, we can plant a version of the truncated cone scaled down by a near any point in $\mathcal{R}^d$ by means of suitable translation and rotation, in such a manner that it lies entirely in W_a^c. Intuitively, this ensures that any point in $\mathcal{R}^d$ cannot have points in W arbitrarily close to it in all directions. We shall later show that this implies that a sequence of iterates approaching W will get pushed out to a shrinking family of such truncated cones sufficiently often. In turn, we also show that this is enough to ensure that the iterates move away from W to one of the A_i, whence they cannot converge to W.

We shall keep α fixed henceforth. Thus we may denote the set $x + aO_{x,a}D_\alpha$ as $D^{x,a}$. Let $\hat{I}_d$ denote the d-dimensional identity matrix and let $M_1 \geq M_2$ for a pair of $d \times d$ positive definite matrices stand for $x^\mathsf{T} M_1 x \geq x^\mathsf{T} M_2 x \ \forall x \in \mathcal{R}^d$. The main consequence of (A5) that we shall need is the following:

Lemma 15. *For any $c > b > 0$, there exists a $\beta = \beta(b,c) > 0$ such that for any $a > 0$ sufficiently small, $x \in \mathcal{R}^d$ and any d-dimensional Gaussian measure*

μ *with mean* x *and covariance matrix* Σ *satisfying* $a^2 c \hat{I}_d \geq \Sigma \geq a^2 b \hat{I}_d$, *one has* $\mu(D^{x,a}) \geq \beta(b,c)$.

Proof. By the scaling properties of the Gaussian, $\mu(D^{x,a}) = \hat{\mu}(D^{0,1})$ where $\hat{\mu}$ denotes the Gaussian measure with zero mean and covariance matrix $\hat{\Sigma}$ satisfying

$$c \hat{I}_d \geq \hat{\Sigma} \geq b \hat{I}_d.$$

The claim follows.   ∎

We also assume:

(**A6**) There exists a positive definite matrix-valued continuous map $Q : \mathcal{R}^d \to \mathcal{R}^{d \times d}$ such that for all $n \geq 0$, $E[M_{n+1} M_{n+1}^T | \mathcal{F}_n] = Q(x_n)$ and for some $0 < \Lambda^- < \Lambda^+ < \infty$, $\Lambda^+ \hat{I}_d \geq Q(x) \geq \Lambda^- \hat{I}_d$.

(**A7**) $\sup_n \frac{b(n)}{a(n)} < \infty$.

(**A8**) $h(\cdot)$ is continuously differentiable and the Jacobian matrix $\nabla h(\cdot)$ is locally Lipschitz.

Assumption (A6) intuitively means that the noise is 'rich' enough in all directions. Assumption (A7) is satisfied, e.g., by $a(n) = 1/n$, $a(n) = 1/(1 + n \ell n(n))$, etc., but not by, say, $a(n) = 1/n^{\frac{2}{3}}$. Thus it requires $a(n)$ to decrease 'sufficiently fast'. (We shall mention a possible relaxation of this condition later.) Let $s, T > 0$. Consider a trajectory $x(\cdot)$ of (4.1.2) with $x(s) \in U$, where U is the closure of a bounded open set containing $A \overset{\text{def}}{=} \bigcup_i A_i$. For $t > s$ in $[s, s + T + 1]$, let $\Phi(t,s)$ denote the $\mathcal{R}^{d \times d}$-valued solution of the linear system

$$\dot{\Phi}(t,s) = \nabla h(x(t)) \Phi(t,s), \quad \Phi(s,s) = \hat{I}_d. \qquad (4.3.1)$$

For a positive definite matrix Q, let $\lambda_{min}(Q), \lambda_{max}(Q)$ denote the least and the highest eigenvalue of Q. Let

$$c^* \overset{\text{def}}{=} \sup \lambda_{max}(\Phi(t,s)\Phi^T(t,s)),$$

$$b^* \overset{\text{def}}{=} \inf \lambda_{min}(\Phi(t,s)\Phi^T(t,s)),$$

where the superscript 'T' denotes matrix transpose and the supremum and infimum are over all $x(\cdot)$ as above *and* all $s + T + 1 \geq t \geq s \geq 0$. Then $\infty > c^* \geq b^* > 0$. The leftmost inequality follows from the fact that $\nabla h(x)$ is uniformly bounded because of the Lipschitz condition on h, whence a standard argument using the Gronwall inequality implies a uniform upper bound on $\|\Phi(t,s)\|$ for t, s in the above range. The rightmost inequality, on the other hand, is a consequence of the fact that $\Phi(t, s)$ is nonsingular for all $t > s$ in the above range. Also, the time dependence of its dynamics is via the continuous dependence of its coefficients on $x(\cdot)$, which lies in a compact set. Hence the smallest eigenvalue of $\Phi(t, s)\Phi^T(t, s)$, being a continuous function of its entries, is bounded away from zero.

For $j \geq m(n), n \geq 0$, let $y_j \overset{\text{def}}{=} x_j - x^n(t(j))$, where $x^n(\cdot)$ is the solution of the o.d.e. (4.1.2) on $[t(n), \infty)$ with $x^n(t(n)) = \bar{x}(t(n))$. Recall that

$$x^n(t(j+1)) = x^n(t(j)) + a(j)h(x^n(t(j))) + O(a(j)^2).$$

Subtracting this from (4.1.1) and using Taylor expansion, one has

$$y_{j+1} = y_j + a(j)(\nabla h(x^n(t(j)))y_j + \kappa_j) + a(j)M_{j+1} + O(a(j)^2),$$

where $\kappa_j = o(\|y_j\|)$. In particular, iterating the expression above leads to

$$
\begin{aligned}
y_{m(n)+i} &= \Pi_{j=m(n)}^{m(n)+i-1}(1 + a(j)\nabla h(x^n(t(j))))y_{m(n)} \\
&\quad + \sum_{j=m(n)}^{m(n)+i-1} a(j)\Pi_{k=j+1}^{m(n)+i-1}(1 + a(k)\nabla h(x^n(t(k))))\kappa_j \\
&\quad + \sum_{j=m(n)}^{m(n)+i-1} a(j)\Pi_{k=j+1}^{m(n)+i-1}(1 + a(k)\nabla h(x^n(t(k))))M_{j+1} \\
&\quad + O(a(m(n))).
\end{aligned}
$$

Since $y_{m(n)} = 0$, the first term drops out. The second term tends to zero as $n \uparrow \infty$ because $\|y_n\|$ does. The last term clearly tends to zero as $n \uparrow \infty$. Let Ψ_n denote the third term on the right when $i = m(n+1) - m(n)$, and let $\hat{\Psi}_n \overset{\text{def}}{=} \Psi_n/\varphi(n)$, where

$$\varphi(n) \overset{\text{def}}{=} (b(m(n)) - b(m(n+1)))^{1/2}.$$

Let $\mathcal{P}(\mathcal{R}^d)$ denote the space of probability measures on $\mathcal{R}^d$ with Prohorov topology (see Appendix C). The next lemma is a technical result which we need later. Let ϕ_n denote the regular conditional law of $\hat{\Psi}_n$ given $\mathcal{F}_{m(n)}, n \geq 0$, viewed as a $\mathcal{P}(\mathcal{R}^d)$-valued random variable.

Lemma 16. *Almost surely on* $\{\bar{x}(t(m(n))) \in U \ \forall n\}$, *every limit point of* $\{\phi_n\}$ *as* $n \uparrow \infty$ *is zero mean Gaussian with the spectrum of its covariance matrix contained in* $[b^*\Lambda^-, c^*\Lambda^+]$.

Proof. For $n \geq 0$ define the martingale array $\{\xi_i^n, 0 \leq i \leq k_n \overset{\text{def}}{=} m(n+1) - m(n)\}$ by $\xi_0^n = 0$ and

$$\xi_i^n = \frac{1}{\varphi(n)} \sum_{j=m(n)}^{m(n)+i-1} a(j)\Pi_{k=j+1}^{m(n)+i-1}(1 + a(k)\nabla h(x^n(t(k)))) M_{j+1}.$$

Then $\hat{\Psi}_n = \xi^n_{k_n}$ and if $\langle \xi^n \rangle_i, 0 \le i \le k_n$, denotes the corresponding matrix-valued quadratic covariation process, i.e.,

$$\langle \xi^n \rangle_m \stackrel{\text{def}}{=} \sum_{i=0}^{m} E[(\xi^n_{i+1} - \xi^n_i)(\xi^n_{i+1} - \xi^n_i)^{\mathrm{T}} | \mathcal{F}_i],$$

then

$$\langle \xi^n \rangle_m = \frac{1}{\varphi(n)^2} \sum_{j=m(n)}^{m(n)+i-1} a(j)^2 \left(\Pi^{m(n)+i-1}_{k=j+1}(1 + a(k)\nabla h(x^n(t(k)))) \right)$$

$$\times Q(x_{m(n)+j}) \left(\Pi^{m(n)+i-1}_{k=j+1}(1 + a(k)\nabla h(x^n(t(k)))) \right)^{\mathrm{T}}.$$

As $n \uparrow \infty$,

$$\frac{1}{\varphi(n)^2} \sum_{j=m(n)}^{m(n)+i-1} a(j)^2 \left(\Pi^{m(n)+i-1}_{k=j+1}(1 + a(k)\nabla h(x^n(t(k)))) \right)$$

$$\times Q(x_{m(n)+j}) \left(\Pi^{m(n)+i-1}_{k=j+1}(1 + a(k)\nabla h(x^n(t(k)))) \right)^{\mathrm{T}}$$

$$- \frac{1}{\varphi(n)^2} \sum_{j=m(n)}^{m(n)+i-1} a(j)^2 \left(\Pi^{m(n)+i-1}_{k=j+1}(1 + a(k)\nabla h(x^n(t(k)))) \right)$$

$$\times Q(x^n(t(m(n)+j))) \left(\Pi^{m(n)+i-1}_{k=j+1}(1 + a(k)\nabla h(x^n(t(k)))) \right)^{\mathrm{T}}$$

$$\rightarrow 0, \quad \text{a.s.}$$

by Lemma 1 and Theorem 2 of Chapter 2. Note that $x^n(\cdot)$ is $\mathcal{F}_{m(n)}$-measurable. Fix a sample point in the probability one set where the conclusions of Lemma 1 and Theorem 2 of Chapter 2 hold. Pick a subsequence $\{n(\ell)\} \subset \{n\}$ such that $x_{m(n(\ell))} \rightarrow x^*$ (say). Then $x^{n(\ell)}(\cdot) \rightarrow \tilde{x}(\cdot)$ uniformly on compact intervals, where $\tilde{x}(\cdot)$ is the unique solution to the o.d.e. $\dot{\tilde{x}}(t) = h(\tilde{x}(t))$ with $\tilde{x}(0) = x^*$. Along $\{n(\ell)\}$, any limit point of the matrices

$$\left(\Pi^{m(n)+i-1}_{k=j+1}(1 + a(k)\nabla h(x^n(t(k)))) \right) Q(x^n(t(m(n)+j)))$$

$$\times \left(\Pi^{m(n)+i-1}_{k=j+1}(1 + a(k)\nabla h(x^n(t(k)))) \right)^{\mathrm{T}}, \quad m(n) \le j < m(n)+i, \ i \ge 0,$$

is of the form $\Phi(t,s)Q(x)\Phi(t,s)^{\mathrm{T}}$ for some t, s, x and therefore has its spectrum in $[b^*\Lambda^-, c^*\Lambda^+]$. Hence the same is true for any convex combinations or limits of convex combinations thereof. In view of this, the claim follows on applying the central limit theorem for martingale arrays (Chow and Teicher, 2003, p. 351; see also Hall and Heyde, 1980) to $\{\phi_{n(j)}\}$. ∎

Remark: The central limit theorem for martingale arrays referred to above is stated in Chow and Teicher (2003) for the scalar case, but the vector case is easily deducible from it by applying the scalar case to arbitrary one-dimensional projections thereof.

Clearly $x_n \to W \cup (\cup_i A_i)$, because any internally chain recurrent invariant set must be contained in $W \cup (\cup_i A_i)$. But Theorem 2 of Chapter 2 implies the connectedness of the a.s. limit set of $\{x_n\}$, and W and $\cup_i A_i$ have disjoint open neighbourhoods. Thus it follows that the sets $\{x_n \to W\}$ and $\{x_n \to W$ along a subsequence$\}$ are a.s. identical. We also need:

Lemma 17. *If F_n, H_n are events in $\mathcal{F}_n$, $n \geq 0$, such that $P(F_{n+1}|\mathcal{F}_n) \geq \kappa > 0$ on H_n for all $n \geq 0$, then $P(\{F_n \ i.o.\}^c \bigcap \{H_n \ i.o.\}) = 0$.*

Proof. Since

$$Z_n \stackrel{\text{def}}{=} \sum_{m=0}^{n-1} I_{F_{m+1}} - \sum_{m=0}^{n-1} P(F_{m+1}|\mathcal{F}_m), \ n \geq 0, \qquad (4.3.2)$$

is a zero mean martingale with bounded increments, almost surely it either converges or satisfies

$$\limsup_{n \to \infty} Z_n = -\liminf_{n \to \infty} Z_n = \infty.$$

(See Theorem 12 of Appendix C.) Thus the two sums on the right-hand side of (4.3.2) converge or diverge together, a.s. Since the second sum is larger than $\kappa \sum_{m=0}^{n-1} I_{H_m}$, it follows that $\sum_{m=0}^{n} I_{F_n}, n \geq 0$, diverges a.s. whenever $\sum_{m=0}^{n} I_{H_m}, n \geq 0$, does. The claim follows. ∎

Recall that U is the closure of a bounded open neighbourhood of $\cup_i A_i$.

Corollary 18. *For any $r > 0$, $\{x_{m(n+1)} \in W^c_{r\varphi(n)} \ i.o.\}$ a.s. on the set $\{x_n \to W\}$.*

Proof. By Lemmas 15 and 16, it follows that almost surely for n sufficiently large, the conditional probability

$$P(x_{m(n+1)} \in W^c_{r\varphi(n)}|\mathcal{F}_n)$$

satisfies

$$P(x_{m(n+1)} \in W^c_{r\varphi(n)}|\mathcal{F}_n) > \eta > 0$$

on the set $\{x_{m(n)} \in U\}$, for some η independent of n. It follows from Lemma 17 that $\{x_{m(n+1)} \in W^c_{r\varphi(n)} \ i.o.\}$ a.s. on the set $\{x_n \in U$ from some n on and $x_n \to W\}$. The claim follows by applying this to a countable increasing family of sets U that covers $\mathcal{R}^d$. ∎

We now return to the framework of the preceding section with $B = U \cap W^c$. Recall our choice of n_0 such that (4.1.4) holds. Let $\hat{K}$ be a prescribed positive constant. By (A7),

$$\sup_n \frac{b(m(n))}{b(m(n)) - b(m(n+1))} = \sup_n \frac{b(m(n))}{\varphi(n)^2} \leq 1 + \sup_n \frac{b(m(n+1))c}{a(m(n+1))T} < \infty,$$

where we have used the fact that $(b(m(n)) - b(m(n+1))) \geq \frac{Ta(m(n+1))}{c}$. Thus we have

$$\frac{b(m(n))}{\varphi(n)} \to 0 \text{ as } n \uparrow \infty.$$

Thus for $\delta = \hat{K}\varphi(n)$, we do have

$$\tilde{K}b(i) < \frac{\delta}{2}, \ \forall i \geq m(n),$$

for $\tilde{K}$ as in Lemma 7 and Theorem 8 and for n sufficiently large, say $n \geq n_0$. (This can be ensured by increasing n_0 if necessary.) With this choice of δ and for $x_{m(n)} \in B$, the probability that $x_k \to A$ exceeds

$$1 - \frac{\tilde{K}b(m(n))}{\hat{K}^2\varphi(n)^2}. \tag{4.3.3}$$

As noted above, $\sup_n \frac{b(m(n))}{\varphi(n)^2} < \infty$. Thus we may choose $\hat{K}$ large enough that the right-hand side of (4.3.3) exceeds $\frac{1}{2}$, with n sufficiently large. Then we have our main result:

Theorem 19. $x_n \to A$ a.s.

Proof. Take $r = \hat{K}$ in Corollary 18. By the foregoing,

$$P(x_{m(n)+k} \overset{k\uparrow\infty}{\to} A | \mathcal{F}_{m(n)+1}) \geq \frac{1}{2},$$

on the set $\{x_{m(n+1)} \in W^c_{r\varphi(n)} \cap U\}$ for $n \geq 0$ sufficiently large. It follows from Lemma 17 that $x_n \to A$ a.s. on $\{x_{m(n+1)} \in W^c_{r\varphi(n)} \cap U \text{ i.o.}\}$, therefore a.s. on $\{x_{m(n+1)} \in W^c_{r\varphi(n)} \text{ i.o.}\}$ (by considering countably many sets U that cover $\mathcal{R}^d$), and finally, a.s. on $\{x_n \to W\}$ by the above corollary. That is, $x_n \to A$ a.s. on $\{x_n \to W\}$, a contradiction unless $P(x_n \to W) = P(x_n \to W$ along a subsequence) = 0. ∎

Two important generalizations are worth noting:

(i) We have used (A6) only to prove Lemma 16. So the conclusions of the lemma, which stipulate a condition on *cumulative* noise rather than *per iterate* noise as in (A6), will suffice.

(ii) The only important consequence of (A7) used was the fact that the ratio $b(n)/\varphi(n)^2$, $n \geq 1$, remains bounded. This is actually a much weaker requirement than (A7) itself.

To conclude, Theorem 19 is just one example of an 'avoidance of traps' result. There are several other formulations, notably Ljung (1978), Pemantle (1990), Brandière and Duflo (1996), Brandière (1998) and Benaim (1999). See also Fang and Chen (2000) for some related results.

5

Stochastic Recursive Inclusions

5.1 Preliminaries

This chapter considers an important generalization of the basic stochastic approximation scheme of Chapter 2, which we call 'stochastic recursive inclusions'. The idea is to replace the map $h : \mathcal{R}^d \to \mathcal{R}^d$ in the recursion (2.1.1) of Chapter 2 by a *set-valued* map $h : \mathcal{R}^d \to \{\text{subsets of } \mathcal{R}^d\}$, satisfying the following conditions:

(i) For each $x \in \mathcal{R}^d$, $h(x)$ is convex and compact.

(ii) For all $x \in \mathcal{R}^d$,

$$\sup_{y \in h(x)} \|y\| < K(1 + \|x\|) \qquad (5.1.1)$$

for some $K > 0$.

(iii) h is *upper semicontinuous* in the sense that if $x_n \to x$ and $y_n \to y$ with $y_n \in h(x_n)$ for $n \geq 1$, then $y \in h(x)$. (In other words, the *graph* of h, defined as $\{(x, y) : y \in h(x)\}$, is closed.)

See Aubin and Frankowska (1990) for general background on set-valued maps and their calculus. Stochastic recursive inclusion refers to the scheme

$$x_{n+1} = x_n + a(n)[y_n + M_{n+1}], \qquad (5.1.2)$$

where $\{a(n)\}$ are as before, $\{M_n\}$ is a martingale difference sequence w.r.t. the increasing σ-fields $\mathcal{F}_n = \sigma(x_m, y_m, M_m, m \leq n), n \geq 0$, satisfying (A3) of Chapter 2, and finally, $y_n \in h(x_n) \ \forall n$. The requirement that $\{y_n\}$ be in $\{h(x_n)\}$ is the reason for the terminology 'stochastic recursive inclusions'. We shall give several interesting applications of stochastic recursive inclusions in section 5.3, following the convergence analysis of (5.1.2) in the next section.

5.2 The differential inclusion limit

As might be expected, in this chapter the o.d.e. limit (2.1.5) of Chapter 2 gets replaced by a differential inclusion limit

$$\dot{x}(t) \in h(x(t)). \tag{5.2.1}$$

To prove that (5.2.1) is indeed the desired limiting differential inclusion, we proceed as in Chapter 2 to define $t(0) = 0, t(n) = \sum_{m=0}^{n-1} a(m), n \geq 1$. Define $\bar{x}(\cdot)$ as before, i.e., set $\bar{x}(t(n)) = x_n, n \geq 0$, with linear interpolation on each interval $[t(n), t(n+1)]$. Define the piecewise constant function $\bar{y}(t), t \geq 0$, by $\bar{y}(t) = y_n, t \in [t(n), t(n+1)), n \geq 0$. Define $\{\zeta_n\}$ as before. As in Chapter 2, we shall analyze (5.1.2) under the stability assumption

$$\sup_n \|x_n\| < \infty, \text{ a.s.} \tag{5.2.2}$$

For $s \geq 0$, let $x^s(t), t \geq s$, denote the solution to

$$\dot{x}^s(t) = \bar{y}(t), \ x^s(s) = \bar{x}(s).$$

Tests for whether (5.2.2) holds can be stated, e.g., along the lines of section 3.2. We omit the details. The following can now be proved exactly along the lines of Lemma 1 of Chapter 2.

Lemma 1. *For any $T > 0$, $\lim_{s \to \infty} \sup_{t \in [s, s+T]} \|\bar{x}(t) - x^s(t)\| = 0$ a.s.*

By (5.2.2) and condition (ii) on the set-valued map h,

$$\sup\{\|y\| : y \in \bigcup_n h(x_n)\} < \infty, \text{ a.s.} \tag{5.2.3}$$

Thus almost surely, $\{x^s(\cdot), s \geq 0\}$ is an equicontinuous, pointwise bounded family. By the Arzela–Ascoli theorem, it is therefore relatively compact in $C([0, \infty); \mathcal{R}^d)$. By Lemma 1, the same then holds true also for $\{\bar{x}(s+\cdot) : s \geq 0\}$, because if not, there exist $s_n \uparrow \infty$ such that $\bar{x}(s_n + \cdot)$ does not have any limit point in $C([0, \infty); \mathcal{R}^d)$. Then nor does $x^{s_n}(\cdot)$ by the lemma, a contradiction to the relative compactness of the latter.

Theorem 2. *Almost surely, every limit point $x(\cdot)$ of $\{\bar{x}(s + \cdot), s \geq 0\}$ in $C([0, \infty); \mathcal{R}^d)$ as $s \to \infty$ satisfies (5.2.1). That is, it satisfies $x(t) = x(0) + \int_0^t y(s)ds, t \geq 0$, for some measurable $y(\cdot)$ satisfying $y(t) \in h(x(t)) \ \forall t$.*

Proof. Fix $T > 0$. Viewing $\{\bar{y}(s + t), t \in [0, T]\}, s \geq 0$, as a subset of $L_2([0, T]; \mathcal{R}^d)$, it follows from (5.2.3) that it is bounded and hence weakly relatively sequentially compact. (See Appendix A.) Let $s(n) \to \infty$ be a sequence such that $\bar{x}(s(n) + \cdot) \to x(\cdot)$ in $C([0, \infty); \mathcal{R}^d)$ and $\bar{y}(s(n) + \cdot) \to y(\cdot)$ weakly in $L_2([0,T];\mathcal{R}^d)$. Then by Lemma 1, $x^{s(n)}(s(n) + \cdot) \to x(\cdot)$ in $C([0,\infty);\mathcal{R}^d)$. Letting $n \to \infty$ in the equation

$$x^{s(n)}(t) = x^{s(n)}(0) + \int_0^t \bar{y}(s(n) + z)dz, \ t \geq 0,$$

we have $x(t) = x(0) + \int_0^t y(z)dz, t \geq 0$. Since $\bar{y}(s(n) + \cdot) \to y(\cdot)$ weakly in $L_2([0,T];\mathcal{R}^d)$, there exist $\{n(k)\} \subset \{n\}$ such that $n(k) \uparrow \infty$ and

$$\frac{1}{N}\sum_{k=1}^{N} \bar{y}(s(n(k)) + \cdot) \to y(\cdot)$$

strongly in $L_2([0,T];\mathcal{R}^d)$ (see Appendix A). In turn, there exist $\{N(m)\} \subset \{N\}$ such that $N(m) \uparrow \infty$ and

$$\frac{1}{N(m)}\sum_{k=1}^{N(m)} \bar{y}(s(n(k)) + \cdot) \to y(\cdot) \tag{5.2.4}$$

a.e. in $[0,T]$. Fix $t \in [0,T]$ where this holds. Define $[s] \overset{def}{=} \max\{t(n) : t(n) \leq s\}$. Then $\bar{y}(s(n(k)) + t) \in h(\bar{x}([s(n(k)) + t])) \ \forall k$. Since we have $\bar{x}(s(n) + \cdot) \to x(\cdot)$ in $C([0,\infty);\mathcal{R}^d)$ and $t(n+1) - t(n) \to 0$, it follows that

$$\begin{aligned}
\bar{x}([s(n(k)) + t]) &= (\bar{x}([s(n(k)) + t]) - \bar{x}(s(n(k)) + t)) \\
&\quad + (\bar{x}(s(n(k)) + t) - x(t)) + x(t) \\
&\to x(t).
\end{aligned}$$

The upper semicontinuity of the set-valued map h then implies that

$$\bar{y}(s(n(k)) + t) \to h(x(t)).$$

Since $h(x(t))$ is convex compact, it follows from (5.2.4) that $y(t) \in h(x(t))$. Thus $y(t) \in h(x(t))$ a.e., where the qualification 'a.e.' may be dropped by modifying $y(\cdot)$ suitably on a Lebesgue-null set. Since $T > 0$ was arbitrary, the claim follows. ∎

Before we proceed, here's a technical lemma about (5.2.1):

Lemma 3. *The set-valued map $x \in \mathcal{R}^d \to Q_x \subset C([0,\infty);\mathcal{R}^d)$, where $Q_x \overset{def}{=}$ the set of solutions to (5.2.1) with initial condition x, is nonempty compact valued and upper semicontinuous.*

Proof. From (5.1.1), we have that for a solution $x(\cdot)$ of (5.2.1) with a prescribed initial condition x_0,

$$\|x(t)\| \leq \|x_0\| + K' \int_0^t (1 + \|x(s)\|)ds, \ t \geq 0,$$

for a suitable constant $K' > 0$. By the Gronwall inequality, it follows that any solution $x(\cdot)$ of (5.2.1) with a prescribed initial condition x_0 (more generally, initial conditions belonging to a bounded set) remains bounded on $[0, T]$ for each $T > 0$ by a bound that depends only on T. From (5.1.1) and (5.2.1) it then follows that the corresponding $\|\dot{x}(t)\|$ remains bounded on $[0, T]$ with a bound that depends only on T. By the Arzela–Ascoli theorem, it follows that Q_{x_0} is relatively compact in $C([0, \infty); \mathcal{R}^d)$. Set $y(\cdot) = \dot{x}(\cdot)$ and write

$$x(t) = x_0 + \int_0^t y(s)ds, \ t \geq 0.$$

Now argue as in the proof of Theorem 2 to show that any limit point $(\bar{x}(\cdot), \bar{y}(\cdot))$ in $C([0, \infty); \mathcal{R}^d)$ of Q_{x_0} will also satisfy this equation with $\tilde{y}(t) \in h(\tilde{x}(t))$ a.e., proving that Q_{x_0} is closed. Next consider $x_n \to x_\infty$ and $x^n(\cdot) \in Q_{x_n}$ for $n \geq 1$. Since $\{x_n, \ n \geq 1\}$ is in particular a bounded set, argue as above to conclude that $\{x^n(\cdot), \ n \geq 1\}$ is relatively compact in $C([0, \infty); \mathcal{R}^d)$. An argument similar to that used to prove Theorem 2 can be used once more, to show that any limit point is in Q_{x_∞}. This proves the upper semicontinuity of the map $x \in \mathcal{R}^d \to Q_x \subset C([0, \infty); \mathcal{R}^d)$. ∎

The next result uses the obvious generalizations of the notions of invariant set and chain transitivity to differential inclusions. We shall say that a set B is invariant (resp. positively / negatively invariant) under (5.2.1) if for $x \in B$, there is some trajectory $x(t), t \in (-\infty, \infty)$ (resp. $[0, \infty)$ / $(-\infty, 0]$), that lies entirely in B. Note that we do not require this of *all* trajectories of (5.2.1) passing through x at time 0. That requirement would define a stronger notion of invariance that we shall not be using here. See Benaim, Hofbauer and Sorin (2003) for various notions of invariance for differential inclusions.

Corollary 4. *Under (5.2.2), $\{x_n\}$ generated by (5.1.2) converge a.s. to a closed connected internally chain transitive invariant set of (5.2.1).*

Proof. From the foregoing, we know that $\{x_n\}$ will a.s. converge to $A \stackrel{\text{def}}{=} \cap_{t \geq 0} \overline{\{\bar{x}(t+s) : s \geq 0\}}$. The proof that A is invariant and that for any $\delta > 0$, $\bar{x}(t + \cdot)$ is the the open δ-neighbourhood A^δ of A for t sufficiently large is similar to that of Theorem 2 of Chapter 2 where similar claims are established. It is therefore omitted. To prove internal chain transitivity of A, let $\tilde{x}_1, \tilde{x}_2 \in A$ and $\epsilon, T > 0$. Pick $\epsilon/4 > \delta > 0$. Pick $n_0 > 1$ such that $n \geq n_0$ implies that for $s \geq t(n)$, $\bar{x}(s + \cdot) \in A^\delta$ and furthermore,

$$\sup_{t \in [s, s+2T]} \|\bar{x}(t) - \check{x}^s(t)\| < \delta$$

for some solution $\check{x}^s(\cdot)$ of (5.2.1) in A. Pick $n_2 > n_1 \geq n_0$ such that $\|\bar{x}(t(n_i)) - \tilde{x}_i\| < \delta, i = 1, 2$. Let $kT \leq t(n_2) - t(n_1) < (k + 1)T$ for some integer $k \geq 0$. Let $s(0) = t(n_1), s(i) = s(0) + iT$ for $1 \leq i < k$, and $s(k) = t(n_2)$. Then for $0 \leq i < k$, $\sup_{t \in [s(i), s(i+1)]} \|\bar{x}(t) - \check{x}^{s(i)}(t)\| < \delta$. Pick $\hat{x}_i, 0 \leq i \leq k$, in G such that $\hat{x}_1 = \bar{x}_1$, $\hat{x}_k = \bar{x}_2$, and for $0 < i < k$, $\hat{x}_i$ are in the δ-neighbourhood of $\bar{x}(s(i))$. The sequence $(s(i), \hat{x}_i), 0 \leq i \leq k$, satisfies the definition of internal chain transitivity. ∎

The invariance of A is in fact implied by its internal chain transitivity as shown in Benaim, Hofbauer and Sorin (2003), so it need not be separately established.

5.3 Applications

In this section we consider four applications of the foregoing. A fifth important one is separately dealt with in the next section. We start with a useful technical lemma. Let $\bar{co}(\cdots)$ stand for '*the closed convex hull of* $\cdots$'.

Lemma 5. *Let* $f : x \in \mathcal{R}^d \to f(x) \subset \mathcal{R}^d$ *be an upper semicontinuous set-valued map such that* $f(x)$ *is compact for all* x *and* $\sup\{\|f(x)\| : \|x\| < M\}$ *is bounded for all* $M > 0$. *Then the set-valued map* $x \to \bar{co}(f(x))$ *is also upper semicontinuous.*

Proof. Let $x_n \to x, y_n \to y$, in $\mathcal{R}^d$ such that $y_n \in \bar{co}(f(x_n))$ for $n \geq 1$. Then by Caratheodory's theorem (Theorem 17.1, p. 155, of Rockafellar, 1970), there exist $a_0^n, \dots a_d^n \in [0,1]$ with $\sum_i a_i^n = 1$, and $y_0^n, \dots, y_d^n \in f(x_n)$, not necessarily distinct, so that $\|y_n - \sum_{i=0}^d a_i^n y_i^n\| < \frac{1}{n}$ for $n \geq 1$. By dropping to a suitable subsequence, we may suppose that $a_i^n \to a_i \in [0,1]$ $\forall i$ and $y_i^n \to \hat{y}_i \in f(x)$ $\forall i$. (Here we use the hypotheses that f is upper semicontinuous and bounded on compacts.) Then $\sum_i a_i = 1$ and $y = \sum_i a_i \hat{y}_i \in \bar{co}(f(x))$, which proves the claim. ∎

We now list four instances of stochastic recursive inclusions.

(i) *Controlled stochastic approximation:* Consider the iteration

$$x_{n+1} = x_n + a(n)[g(x_n, u_n) + M_{n+1}],$$

where $\{u_n\}$ is a random sequence taking values in a compact metric space U, $\{a(n)\}$ are as before, $\{M_n\}$ satisfies the usual conditions w.r.t. the σ-fields $\mathcal{F}_n \overset{\text{def}}{=} \sigma(x_m, u_m, M_m, m \leq n), n \geq 0$, and $g : \mathcal{R}^d \times U \to \mathcal{R}^d$ is continuous and Lipschitz in the first argument uniformly w.r.t. the second. We view $\{u_n\}$ as a control process. That is, u_n is chosen by the agent running the algorithm at time $n \geq 0$ based on the observed history and possibly extraneous independent randomization, as in the usual stochastic control problems. It could, however, also be an unknown random process in addition to $\{M_n\}$ that affects the measurements. The idea then is to analyze the asymptotic behaviour of the iterations for arbitrary $\{u_n\}$ that fit the above description. It then makes sense to define $h(x) = \bar{co}(\{g(x,u) : u \in U\})$. The above iteration then becomes a special case of (1). The three conditions stipulated for the set-valued map are easily verified in this particular case by using Lemma 5.

(ii) *Stochastic subgradient descent:* Consider a continuously differentiable convex function $f : \mathcal{R}^d \to \mathcal{R}$ which one aims to minimize based on noisy measurements of its gradients. That is, at any point $x \in \mathcal{R}^d$, one can measure $\nabla f(x) +$ an independent copy of a zero mean random variable. Then the natural scheme to explore would be

$$x_{n+1} = x_n + a(n)[-\nabla f(x_n) + M_{n+1}],$$

where $\{M_n\}$ is the i.i.d. (more generally, a martingale difference) measurement noise and the expression in square brackets on the right represents the noisy measurement of the gradient. This is a special case of the 'stochastic gradient scheme' we shall discuss in much more detail later in the book. Here we are interested in an extension of the scheme that one encounters when f is not continuously differentiable everywhere, so that ∇f is not defined at all points. It turns out that a natural generalization of ∇f to this 'non-smooth' case is the *subdifferential* ∂f. The subdifferential $\partial f(x)$ at x is the set of all y satisfying

$$f(z) \geq f(x) + \langle y, z - x \rangle$$

for all $z \in \mathcal{R}^d$. It is clear that it will be a closed convex set. It can also be shown to be compact nonempty (Theorem 23.4, p. 217, of Rockafellar, 1970) and upper semicontinuous as a function of x. Assume the linear growth property stipulated in (5.1.1) for $h = -\partial f$. Thus we may replace the above stochastic gradient scheme by equation (5.1.2) with this specific choice of h, which yields the stochastic subgradient descent.

Note that the closed convex set of minimizers of f, if nonempty, is contained in $\mathcal{M} = \{x \in \mathcal{R}^d : \theta \in \partial f\}$, where θ denotes the zero vector in $\mathcal{R}^d$. It is then easy to see that at any point on the trajectory of (5.2.1) lying outside $\mathcal{M}$, $f(x(t))$ must be strictly decreasing. Thus any invariant set for (5.2.1) must be contained in $\mathcal{M}$. Corollary 4 then implies that $x_n \to \mathcal{M}$ a.s.

(iii) *Approximate drift:* We may refer to the function h on the right-hand side of (2.1.1), Chapter 2, as the 'drift'. In many cases, there is a desired drift h that we wish to implement, but we have at hand only an approximation of it. One common situation is when the negative gradient in the stochastic gradient scheme, i.e., $h = -\nabla f$ for some continuously differentiable $f : \mathcal{R}^d \to \mathcal{R}$, is not explicitly available but is known only approximately (e.g., as a finite difference approximation).

In such cases, the actual iteration being implemented is

$$x_{n+1} = x_n + a(n)[h(x_n) + \eta_n + M_{n+1}],$$

where η_n is an error term. Suppose the only available information about $\{\eta_n\}$ is that $\|\eta_n\| \leq \epsilon \ \forall n$ for a known $\epsilon > 0$. In this case, one may analyze the iteration as a stochastic recursive inclusion (5.1.2) with

$$
\begin{aligned}
y_n \in \hat{h}(x_n) &\overset{\text{def}}{=} h(x_n) + \overline{B(\epsilon)} \\
&= \{y : \|h(x_n) - y\| \leq \epsilon\}.
\end{aligned}
$$

Here $\overline{B(\epsilon)}$ is the closed ϵ-ball centered at the origin. An important special case when one can analyze its asymptotic behaviour to some extent is the case when there exists a globally asymptotically stable attractor H for the associated o.d.e. (2.1.5) of Chapter 2.

Theorem 6. *Under (5.2.2), given any $\delta > 0$, there exists an $\epsilon_0 > 0$ such that for all $\epsilon \in (0, \epsilon_0)$, $\{x_n\}$ above converge a.s. to the δ-neighbourhood of H.*

Proof. For $\gamma > 0$, define

$$H^\gamma \overset{\text{def}}{=} \{x \in \mathcal{R}^d : \inf_{y \in H} \|x - y\| < \gamma\}.$$

Fix a sample path where (5.2.2) and Lemma 1 hold. Pick $T > 0$ large enough that for any solution $x(\cdot)$ of the o.d.e.

$$\dot{x}(t) = h(x(t))$$

for which $\|x(0)\| \leq C \overset{\text{def}}{=} \sup_n \|x_n\|$, we have $x(t) \in H^{\delta/3}$ for $t \geq T$. Let $x^s(\cdot)$ be as before and let $\hat{x}^s(\cdot)$ denote the solution of the above o.d.e. for $t \geq s$ with $\hat{x}^s(s) = x^s(s)$. Then a simple application of the Gronwall inequality shows that for $\epsilon_0 > 0$ sufficiently small, $\epsilon \leq \epsilon_0$ implies

$$\sup_{t \in [s, s+T]} \|x^s(t) - \hat{x}^s(t)\| < \delta/3.$$

In particular, $x^s(T) \in H^{2\delta/3}$. Hence, since $s > 0$ was arbitrary, it follows by Lemma 1 that for sufficiently large s, $\bar{x}(s + T) \in H^\delta$, i.e., for sufficiently large s, $\bar{x}(s) \in H^\delta$. ∎

(iv) *Discontinuous dynamics:* Consider

$$x_{n+1} = x_n + a(n)[g(x_n) + M_{n+1}],$$

where $g : \mathcal{R}^d \to \mathcal{R}^d$ is merely measurable but satisfies a linear growth condition: $\|g(x)\| \leq K(1 + \|x\|)$ for some $K > 0$. Define $h(x) \overset{\text{def}}{=} \bigcap_{\epsilon > 0} \bar{co}(\{g(y) : \|y - x\| < \epsilon\})$. Then the above iteration may be viewed as a special case of (5.1.2). The three properties stipulated for h above can be verified in a straightforward manner. Note that the differential inclusion limit in this case is one of the standard solution concepts for differential equations with discontinuous right hand sides – see, e.g., p. 50, Filippov (1988).

In fact, *continuous* g which is not Lipschitz can be viewed as a special case of stochastic recursive inclusions. Here $h(x)$ is always the singleton $\{g(x)\}$ and (5.2.1) reduces to an o.d.e. $\dot{x}(t) = g(x(t))$. The catch is that in absence of the Lipschitz condition, the existence of a solution to this o.d.e. is guaranteed, but not its uniqueness, which may fail at *all* points (see, e.g., Chapter II of Hartman, 1982). Thus the weaker claims above with the suitably weakened notion of invariant sets apply in place of the results of Chapter 2.

As for the invariant sets of (5.2.1), it is often possible to characterize them using a counterpart for differential inclusions of the Liapunov function approach, as in Corollary 3 of Chapter 2. See Chapter 6 of Aubin and Cellina (1980) for details.

5.4 Projected stochastic approximation

These are stochastic approximation iterations that are forced to remain in some bounded set G by being projected back to G whenever they go out of G. This avoids the stability issue altogether, now that the iterates are forced to remain in a bounded set. But it can lead to other complications as we see below, so some care is needed in using this scheme.

Thus iteration (2.1.1) of Chapter 2 is replaced by

$$x_{n+1} = \Gamma(x_n + a(n)[h(x_n) + M_{n+1}]), \qquad (5.4.1)$$

where $\Gamma(\cdot)$ is a projection to a prescribed compact set G. That is, $\Gamma = $ the identity map for points in the interior of G, and maps a point outside G to the point in G closest to it w.r.t. the Euclidean distance. (Sometimes some other equivalent metric may be more convenient, e.g., the max-norm $||x||_\infty \stackrel{\text{def}}{=} \max_i |x_i|$.) The map Γ need not be single-valued in general, but it is when G is convex. This is usually the case in practice. Even otherwise, as long as the boundary ∂G of G is reasonably well-behaved, Γ will be single-valued for points outside G that are sufficiently close to ∂G. Again, this is indeed usually the case for our algorithm because our stepsizes $a(n)$ are small, at least for large n, and thus the iteration cannot move from a point inside G to a point far from G in a single step. Hence assuming that Γ is single-valued is not a serious restriction.

First consider the simple case when ∂G is smooth and Γ is Frechet differentiable, i.e., there exists a linear map $\bar{\Gamma}_x(\cdot)$ such that the limit

$$\gamma(x;y) \overset{\text{def}}{=} \lim_{\delta \downarrow 0} \frac{\Gamma(x + \delta y) - x}{\delta}$$

exists and equals $\bar{\Gamma}_x(y)$. (This will be the identity map for x in the interior of G.) In this case (5.4.1) may be rewritten as

$$\begin{aligned}
x_{n+1} &= x_n + a(n) \frac{\Gamma(x_n + a(n)[h(x_n) + M_{n+1}]) - x_n}{a(n)} \\
&= x_n + a(n)[\bar{\Gamma}_{x_n}(h(x_n)) + \bar{\Gamma}_{x_n}(M_{n+1}) + o(a(n))].
\end{aligned}$$

This iteration is similar to the original stochastic approximation scheme (2.1.1) with $h(x_n)$ and M_{n+1} replaced resp. by $\bar{\Gamma}_{x_n}(h(x_n))$ and $\bar{\Gamma}_{x_n}(M_{n+1})$, with an additional error term $o(a(n))$. Suppose the map $x \to \bar{\Gamma}_x(h(x))$ is Lipschitz. If we mimic the proofs of Lemma 1 and Theorem 2 of Chapter 2, the $o(a(n))$ term will be seen to contribute an additional error term of order $o(a(n)T)$ to the bound on $\sup_{t \in [s,s+T]} \|\bar{x}(t) - x^s(t)\|$, where n is such that $t(n) = [s]$. This error term tends to zero as $s \to \infty$. Thus the same proof as before establishes that the conclusions of Theorem 2 of Chapter 2 continue to hold, but with the o.d.e. (2.1.5) of Chapter 2 replaced by the o.d.e.

$$\dot{x}(t) = \bar{\Gamma}_{x(t)}(h(x(t))). \tag{5.4.2}$$

If $x \to \bar{\Gamma}_x(h(x))$ is merely continuous, then the o.d.e. (5.4.2) will have possibly non-unique solutions for any initial condition and the set of solutions as a function of initial condition will be a compact-valued upper semicontinuous set-valued map. An analog of Theorem 2 of Chapter 2 can still be established with the weaker notion of invariance introduced just before Corollary 4 above. If the map is merely measurable, we are reduced to the 'discontinuous dynamics' scenario discussed above.

Unlike the convexity of G, the requirement that ∂G be smooth is, however, a serious restriction. This is because it fails in many simple cases such as when G is a polytope, which is a very common situation. Thus there is a need to extend the foregoing to cover such cases. This is where the developments of section 5.3 come into the picture.

This analysis may fail in two ways. The first is that the limit $\gamma(x;y)$ above may be undefined. In most cases arising in applications, however, this is not the problem. The difficulty usually is that the limit does exist but does not correspond to the evaluation of a linear map $\bar{\Gamma}_x$ as stipulated above. That is, $\gamma(x;y)$ exists as a *directional derivative* of Γ at x in the direction y, but Γ is not Frechet differentiable. Thus we have the iteration

$$x_{n+1} = x_n + a(n)[\gamma(x_n; h(x_n) + M_{n+1}) + o(a(n))]. \tag{5.4.3}$$

There is still some hope of an o.d.e. limit if M_{n+1} is conditionally independent of $\mathcal{F}_n$ given x_n. In this case, let its regular conditional law given x_n be denoted by $\mu(x, dy)$. Then the above may be rewritten as

$$x_{n+1} = x_n + a(n)[\bar{h}(x_n) + \bar{M}_{n+1} + o(a(n))],$$

where

$$\bar{h}(x_n) \stackrel{\text{def}}{=} \int \mu(x_n, dy)\gamma(x_n; h(x_n) + y),$$

and

$$\bar{M}_{n+1} \stackrel{\text{def}}{=} \gamma(x_n; h(x_n) + M_{n+1}) - \bar{h}(x_n).$$

To obtain this, we have simply added and subtracted on the right-hand side of (5.4.3) the one-step conditional expectation of $\gamma(x_n; h(x_n) + M_{n+1})$. The advantage is that the present expression is in the same format as (2.2.1) modulo the $o(a(n))$ term whose contribution is asymptotically negligible. Thus in the special case when $\bar{h}$ turns out to be Lipschitz, one has the o.d.e. limit

$$\dot{x}(t) = \bar{h}(x(t)),$$

with the associated counterpart of Theorem 2 of Chapter 2.

This situation, however, is very special. Usually $\bar{h}$ is only measurable. Then this reduces to the case of 'discontinuous dynamics' studied in the preceding section and can be analyzed in that framework by means of an appropriate limiting differential inclusion.

In the case when one is saddled with only (5.4.3) and nothing more, let

$$y_n \stackrel{\text{def}}{=} E[\gamma(x_n; h(x_n) + M_{n+1})|\mathcal{F}_n]$$

and

$$\tilde{M}_{n+1} \stackrel{\text{def}}{=} (\gamma(x_n; h(x_n) + M_{n+1}) - y_n)$$

for $n \geq 0$. We then recover (5.1.2) above with $\tilde{M}_{n+1}$ replacing M_{n+1}. Suppose there exists a set-valued map, denoted by $x \to \hat{\Gamma}_x(h(x))$ to suggest a kinship with $\bar{\Gamma}$ above, such that

$$y_n \in \hat{\Gamma}_{x_n}(h(x_n)) \text{ a.s.},$$

and suppose that it satisfies the three conditions stipulated in section 5.1, viz., it is compact convex valued and upper semicontinuous with bounded range on compacts such that (5.1.1) holds. Then the analysis of section 5.1 applies with the limiting differential inclusion

$$\dot{x}(t) \in \hat{\Gamma}_{x(t)}(h(x(t))). \tag{5.4.4}$$

For example, if the support of the conditional distribution of M_{n+1} given $\mathcal{F}_n$ is a closed bounded set $A(x_n)$ depending on x_n, one may consider

$$\hat{\Gamma}_x(h(x)) \stackrel{\text{def}}{=} \bigcap_{\epsilon > 0} \bar{co}\Big(\bigcup_{||z-x|| < \epsilon} \{ \gamma(z; h(z) + y) : y \in A(z) \} \Big).$$

In many situations, ∂G is smooth except along a 'thin' set consisting of a union of surfaces (submanifolds) one or more dimensions lower than ∂G itself. (This is the case, e.g., when G is a polytope, when ∂G is a union of its faces and is nonsmooth at the boundaries of these faces.) If h and the noise $\{M_n\}$ are such that these parts of the boundary 'repel' the iterates the way unstable equilibria were seen to do in Chapter 4, then one can still work with the limiting o.d.e. (5.4.2), ignoring the region where it is not justified. Another scenario when things simplify is when the offending part of ∂G is 'thin' in the above sense and everywhere along this set the possible directions stipulated by the right-hand side of (5.4.4) are such that any solution of (5.4.4) spends zero net time in this set. In this case, (5.4.4) becomes the same as (5.4.2) interpreted in the Caratheodory sense (see pp. 3–4 of Filippov, 1988).

The second major concern in projected algorithms is the possibility of spurious equilibria or other invariant sets on ∂G, i.e., equilibria or invariant sets for (5.4.2) or (5.4.4) that are not equilibria or invariant sets for the o.d.e. (2.1.5). For example, if $h(x)$ is directed along the outward normal at some $x \in \partial G$ and ∂G is smooth in a neighbourhood of x, then x can be a spurious stable equilibrium for the limiting projected o.d.e. These spurious equilibria will be potential asymptotic limit sets for the projected scheme in view of Corollary 4. Thus their presence can lead to convergence of (5.1.2) to undesired points or sets. This has to be avoided where possible by using any prior knowledge available to choose G properly. Another possibility is the following: Suppose we consider a parametrized family of candidate G, say closed balls of radius r centered at the origin. Suppose such problems arise only for r belonging to a Lebesgue-null set. Then we may choose G randomly at each iterate according to some Lebesgue-continuous density for r in a neighbourhood of a nominal value $r = r_0$ fixed beforehand. A further possibility, due to Chen (1994, 1998), is to start with a specific G and slowly increase it to the whole of R^d. We shall revisit this scheme in the next chapter.

It is also possible that the new equilibria or invariant sets on ∂G thus introduced correspond in fact to desired equilibria or invariant sets lying outside G or at ∞. In this case, the former may be viewed as approximations of the latter and thus may in fact be the desired limit sets for the projected algorithm.

In case the limit in the definition of $\gamma(\cdot; \cdot)$ above is not even well-defined, one can consider the set of *all* limit points therein and build a stochastic recursive inclusion around that. We ignore this possibility as it does not seem very useful in applications.

On the flip side, there are situations where the projected dynamics is in fact well-posed, see, e.g., Dupuis and Nagurney (1993).

6

Multiple Timescales

6.1 Two timescales

In the preceding chapters we have used a fixed stepsize schedule $\{a(n)\}$ for all components of the iterations in stochastic approximation. In the 'o.d.e. approach' to the analysis of stochastic approximation, these are viewed as discrete nonuniform time steps. Thus one can conceive of the possibility of using different stepsize schedules for different components of the iteration, which will then induce different timescales into the algorithm. We shall consider the case of two timescales first, following Borkar (1996). Thus we are interested in the iterations

$$x_{n+1} = x_n + a(n)[h(x_n, y_n) + M_{n+1}^{(1)}], \qquad (6.1.1)$$

$$y_{n+1} = y_n + b(n)[g(x_n, y_n) + M_{n+1}^{(2)}], \qquad (6.1.2)$$

where $h : \mathcal{R}^{d+k} \to \mathcal{R}^d, g : \mathcal{R}^{d+k} \to \mathcal{R}^k$ are Lipschitz and $\{M_n^{(1)}\}, \{M_n^{(2)}\}$ are martingale difference sequences w.r.t. the increasing σ-fields

$$\mathcal{F}_n \stackrel{def}{=} \sigma(x_m, y_m, M_m^1, M_m^2, m \leq n), \ n \geq 0,$$

satisfying

$$E[||M_{n+1}^i||^2 | \mathcal{F}_n] \leq K(1 + ||x_n||^2 + ||y_n||^2), \ i = 1, 2,$$

for $n \geq 0$. Stepsizes $\{a(n)\}, \{b(n)\}$ are positive scalars satisfying

$$\sum_n a(n) = \sum_n b(n) = \infty, \quad \sum_n (a(n)^2 + b(n)^2) < \infty, \quad \frac{b(n)}{a(n)} \to 0.$$

The last condition implies that $b(n) \to 0$ at a faster rate than $\{a(n)\}$, implying that (6.1.2) moves on a slower timescale than (6.1.1). Examples of such stepsizes are $a(n) = \frac{1}{n}, b(n) = \frac{1}{1+n\log n}$, or $a(n) = \frac{1}{n^{2/3}}, b(n) = \frac{1}{n}$, and so on.

It is instructive to compare this coupled iteration to the singularly perturbed o.d.e.

$$\dot{x}(t) = \frac{1}{\epsilon} h(x(t), y(t)), \tag{6.1.3}$$

$$\dot{y}(t) = g(x(t), y(t)), \tag{6.1.4}$$

in the limit $\epsilon \downarrow 0$. Thus $x(\cdot)$ is a fast transient and $y(\cdot)$ the slow component. It then makes sense to think of $y(\cdot)$ as quasi-static (i.e., 'almost a constant') while analyzing the behaviour of $x(\cdot)$. This suggests looking at the o.d.e.

$$\dot{x}(t) = h(x(t), y), \tag{6.1.5}$$

where y is held fixed as a constant parameter. Suppose that:

(A1) (6.1.5) has a globally asymptotically stable equilibrium $\lambda(y)$ (*uniformly* in y), where $\lambda : \mathcal{R}^k \to \mathcal{R}^d$ is a Lipschitz map.

Then for sufficiently small values of ϵ we expect $x(t)$ to closely track $\lambda(y(t))$ for $t > 0$. In turn this suggests looking at the o.d.e.

$$\dot{y}(t) = g(\lambda(y(t)), y(t)), \tag{6.1.6}$$

which should capture the behaviour of $y(\cdot)$ in (6.1.4) to a good approximation. Suppose that:

(A2) The o.d.e. (6.1.6) has a globally asymptotically stable equilibrium y^*.

Then we expect $(x(t), y(t))$ in (6.1.3)–(6.1.4) to approximately converge to (i.e., converge to a small neighbourhood of) the point $(\lambda(y^*), y^*)$.

This intuition indeed carries over to the iterations (6.1.1)–(6.1.2). Thus (6.1.1) views (6.1.2) as quasi-static while (6.1.2) views (6.1.1) as almost equilibrated. The motivation for studying this set-up comes from the following considerations. Suppose that an iterative algorithm calls for a particular subroutine in each iteration. Suppose also that this subroutine itself is another iterative algorithm. The traditional method would be to use the output of the subroutine after running it 'long enough' (i.e., until near-convergence) during each iterate of the outer loop. But the foregoing suggests that we could get the same effect by running both the inner and the outer loops (i.e., the corresponding iterations) concurrently, albeit on different timescales. Then the inner 'fast' loop sees the outer 'slow' loop as quasi-static while the latter sees the former as nearly equilibrated. We shall see applications of this later in the book.

We now take up the formal convergence analysis of the two-timescale scheme (6.1.1)–(6.1.2) under the stability assumption:

(A3) $\sup_n(\|x_n\| + \|y_n\|) < \infty$, a.s.

Assume (A1)–(A3) above.

Lemma 1. $(x_n, y_n) \to \{(\lambda(y), y) : y \in \mathcal{R}^k\}$ a.s.

Proof. Rewrite (6.1.2) as

$$y_{n+1} = y_n + a(n)[\epsilon_n + M_{n+1}^{(3)}], \qquad (6.1.7)$$

where $\epsilon_n \overset{\text{def}}{=} \frac{b(n)}{a(n)} g(x_n, y_n)$ and $M_{n+1}^{(3)} \overset{\text{def}}{=} \frac{b(n)}{a(n)} M_{n+1}^{(2)}$ for $n \geq 0$. Consider the pair (6.1.1), (6.1.7) in the framework of the third 'extension' listed at the start of section 2.2. By the observations made there, it then follows that (x_n, y_n) converges to the internally chain transitive invariant sets of the o.d.e. $\dot{x}(t) = h(x(t), y(t)), \dot{y}(t) = 0$. The claim follows. ∎

In other words, $\|x_n - \lambda(y_n)\| \to 0$ a.s., that is, $\{x_n\}$ asymptotically 'track' $\{\lambda(y_n)\}$, a.s.

Theorem 2. $(x_n, y_n) \to (\lambda(y^*), y^*)$ a.s.

Proof. Let $s(0) = 0$ and $s(n) = \sum_{i=0}^{n-1} b(i)$ for $n \geq 1$. Define the piecewise linear continuous function $\tilde{y}(t), t \geq 0$, by $\tilde{y}(s(n)) = y_n$, with linear interpolation on each interval $[s(n), s(n+1)], n \geq 0$. Let $\psi_n \overset{\text{def}}{=} \sum_{m=0}^{n-1} b(m) M_{m+1}^{(2)}, n \geq 1$. Then arguing as for $\{\zeta_n\}$ in Chapter 2, $\{\psi_n\}$ is an a.s. convergent square-integrable martingale. Let $[t]' \overset{\text{def}}{=} \max\{s(n) : s(n) \leq t\}, t \geq 0$. Then for $n, m \geq 0$,

$$
\begin{aligned}
\tilde{y}(s(n+m)) &= \tilde{y}(s(n)) + \int_{s(n)}^{s(n+m)} g(\lambda(\tilde{y}(t)), \tilde{y}(t)) dt \\
&+ \int_{s(n)}^{s(n+m)} (g(\lambda(\tilde{y}([t]')), \tilde{y}([t]')) - g(\lambda(\tilde{y}(t)), \tilde{y}(t))) dt \\
&+ \sum_{k=1}^{m-1} b(n+k)(g(x_{n+k}, y_{n+k}) - g(\lambda(y_{n+k}), y_{n+k})) \\
&+ (\psi_{n+m+1} - \psi_n).
\end{aligned}
$$

For $s \geq 0$, let $y^s(t), t \geq s$, denote the trajectory of (6.1.6) with $y^s(s) = \tilde{y}(s)$. Using the Gronwall inequality as in the proof of Lemma 1 of Chapter 2, we obtain, for $T > 0$,

$$\sup_{t \in [s, s+T]} \|\tilde{y}(t) - y^s(t)\| \leq K_T(I + II + III),$$

where $K_T > 0$ is a constant depending on T and

(i) 'I' is the 'discretization error' contributed by the third term on the right-hand side above, which is $O(\sum_{k \geq n} b(k)^2)$ a.s.,

(ii) '*II*' is the 'error due to noise' contributed by the fifth term on the right-hand side above, which is $O(\sup_{k \geq n} \|\psi_k - \psi_n\|)$ a.s., and

(iii) '*III*' is the 'tracking error' contributed by the fourth term on the right-hand side above, which is $O(\sup_{k \geq n} \|x_k - \lambda(y_k)\|)$ a.s.

Since all three errors tend to zero a.s. as $s \to \infty$,

$$\sup_{t \in [s, s+T]} \|\tilde{y}(t) - y^s(t)\| \to 0, \text{ a.s.}$$

Arguing as in the proof of Theorem 2 of Chapter 2, we get $y_n \to y^*$ a.s. By Lemma 5.1, $x_n \to \lambda(y^*)$ a.s. This completes the proof. ∎

The same general scheme can be extended to three or more timescales. This extension, however, is not as useful as it may seem, because the convergence analysis above captures only the asymptotic 'mean drift' for (6.1.1)–(6.1.2), not the fluctuations about the mean drift. Unless the timescales are reasonably separated, the behaviour of the coupled scheme (6.1.1)–(6.1.2) will not be very graceful. At the same time, if the timescales are greatly separated, that separation may render either the fast timescale too fast (increasing both discretization error and noise-induced error because of larger stepsizes), or the slow timescale too slow (slowing down the convergence because of smaller stepsizes), or both. This difficulty becomes more pronounced the larger the number of timescales involved.

Another, less elegant way of achieving the two-timescale effect would be to run (6.1.2) also with stepsizes $\{a(n)\}$, but along a subsample $\{n(k)\}$ of time instants that become increasingly rare (i.e., $n(k+1) - n(k) \to \infty$) and keeping its values constant between these instants. That is,

$$y_{n(k)+1} = y_{n(k)} + a(n(k))[g(x_{n(k)}, y_{n(k)}) + M^{(2)}_{n(k)+1}],$$

with $y_{n+1} = y_n \; \forall n \notin \{n(k)\}$. In practice it has been found that a good policy is to run (6.1.2) with a slower stepsize schedule $\{b(n)\}$ as above *and* also update it along a subsequence $\{nN, n \geq 0\}$ for a suitable integer $N > 1$, keeping its values constant in between (S. Bhatnagar, personal communication).

6.2 Averaging the natural timescale: preliminaries

Next we consider a situation wherein the stochastic approximation iterations are also affected by another process $\{Y_n\}$ running in the background on the true or 'natural' timescale which corresponds to the time index 'n' itself that tags the iterations. Given our viewpoint of $\{a(n)\}$ as time steps, since $a(n) \to 0$ the algorithm runs on a slower timescale than $\{Y_n\}$ and thus should see the 'averaged' effects of the latter. We make this intuition precise in what follows. This section, which, along with the next section, is based on Borkar (2006), builds up the technical infrastructure for the main results to be presented in the next section. This development requires the background material summarized in Appendix C on spaces of probability measures on metric spaces.

Specifically, we consider the iteration

$$x_{n+1} = x_n + a(n)[h(x_n, Y_n) + M_{n+1}], \qquad (6.2.1)$$

where $\{Y_n\}$ is a random process taking values in a complete separable metric space S with dynamics we shall soon specify, and $h : \mathcal{R}^d \times S \to \mathcal{R}^d$ is jointly continuous in its arguments and Lipschitz in its first argument uniformly w.r.t. the second. $\{M_n\}$ is a martingale difference sequence w.r.t. the σ-fields $\mathcal{F}_n \overset{\text{def}}{=} \sigma(x_m, Y_m, M_m, m \leq n), n \geq 0$. Stepsizes $\{a(n)\}$ are as before, with the additional condition that they be eventually nonincreasing.

We shall assume that $\{Y_n\}$ is an S-valued controlled Markov process with two control processes: $\{x_n\}$ above and another random process $\{Z_n\}$ taking values in a compact metric space U. Thus

$$P(Y_{n+1} \in A | Y_m, Z_m, x_m, m \leq n) = \int_A p(dy | Y_n, Z_n, x_n), \ n \geq 0, \qquad (6.2.2)$$

for A Borel in S, where $(y, z, x) \in S \times U \times \mathcal{R}^d \to p(dw | y, z, x) \in \mathcal{P}(S)$ is a continuous map specifying the controlled transition probability kernel. (Here and in what follows, $\mathcal{P}(\cdots)$ will denote the space of probability measures on the complete separable metric space '$\cdots$' with Prohorov topology – see, e.g., Appendix C.) We assume that the continuity in the x variable is uniform on compacts w.r.t. the other variables. We shall say that $\{Z_n\}$ is a *stationary control* if $Z_n = v(Y_n) \ \forall n$ for some measurable $v : S \to U$, and a *stationary randomized control* if for each n the conditional law of Z_n given $(Y_m, x_m, Z_{m-1}, m \leq n)$ is $\varphi(Y_n)$ for a fixed measurable map $\varphi : y \in S \to \varphi(y) = \varphi(y, dz) \in \mathcal{P}(U)$ independent of n. Thus, in particular, Z_n will then be conditionally independent of $(Y_{m-1}, x_m, Z_{m-1}, m \leq n)$ given Y_n for $n \geq 0$. By abuse of terminology, we identify the stationary (resp. stationary randomized) control above with the map $v(\cdot)$ (resp. $\varphi(\cdot)$). Note that the former is a special case of the latter for $\varphi(\cdot) = \delta_{v(\cdot)}$, where δ_x denotes the Dirac measure at x.

If $x_n = x \ \forall n$ for a fixed deterministic $x \in \mathcal{R}^d$, then $\{Y_n\}$ will be a time-homogeneous Markov process under any stationary randomized control φ. Its transition kernel will be

$$\bar{p}_{x,\varphi}(dw | y) = \int p(dw | y, z, x) \varphi(y, dz).$$

Suppose that this Markov process has a (possibly nonunique) invariant probability measure $\eta_{x,\varphi}(dy) \in \mathcal{P}(S)$. Correspondingly we define the *ergodic occupation measure*

$$\Psi_{x,\varphi}(dy, dz) \overset{\text{def}}{=} \eta_{x,\varphi}(dy) \varphi(y, dz) \in \mathcal{P}(S \times U).$$

This is the stationary law of the state-control pair when the stationary randomized control φ is used and the initial distribution is $\eta_{x,\varphi}$. It clearly satisfies the equation

$$\int_S f(y)d\Psi_{x,\varphi}(dy, U) = \int_S \int_U f(w)p(dw|y, z, x)d\Psi_{x,\varphi}(dy, dz) \qquad (6.2.3)$$

for bounded continuous $f : S \to \mathcal{R}$. Conversely, if some $\Psi \in \mathcal{P}(S \times U)$ satisfies (6.2.3) for f belonging to any set of bounded continuous functions $S \to \mathcal{R}$ that separates points of $\mathcal{P}(S)$, then it must be of the form $\Psi_{x,\varphi}$ for some stationary randomized control φ. (In particular, countable subsets of $C_b(S)$ that separate points of $\mathcal{P}(S)$ are known to exist – see Appendix C.) This is because we can always decompose Ψ as

$$\Psi(dy, dz) = \eta(dy)\varphi(y, dz)$$

with η and φ denoting resp. the marginal on S and the regular conditional law on U. Since $\varphi(\cdot)$ is a measurable map $S \to \mathcal{P}(U)$, it can be identified with a stationary randomized control. (6.2.3) then implies that η is an invariant probability measure under the 'stationary randomized control' φ.

We denote by $D(x)$ the set of all such ergodic occupation measures for the prescribed x. Since (6.2.3) is preserved under convex combinations and convergence in $\mathcal{P}(S \times U)$, $D(x)$ is closed and convex. We also assume that it is compact. Once again, using the fact that (6.2.3) is preserved under convergence in $\mathcal{P}(S \times U)$, it follows that if $x(n) \to x$ in $\mathcal{R}^d$ and $\Psi_n \to \Psi$ in $\mathcal{P}(S \times U)$ with $\Psi_n \in D(x(n)) \; \forall n$, then $\Psi \in D(x)$, implying upper semicontinuity of the set-valued map $x \to D(x)$.

Define $\{t(n)\}, \bar{x}(\cdot)$ as before. We define a $\mathcal{P}(S \times U)$-valued random process $\mu(t) = \mu(t, dydz), t \geq 0$, by

$$\mu(t) \stackrel{\text{def}}{=} \delta_{(Y_n, Z_n)}, \; t \in [t(n), t(n+1)),$$

for $n \geq 0$. This process will play an important role in our analysis of (6.2.1). Also define for $t > s \geq 0$, $\mu_s^t \in \mathcal{P}(S \times U \times [s, t])$ by

$$\mu_s^t(A \times B) \stackrel{\text{def}}{=} \frac{1}{t-s} \int_B \mu(y, A)dy$$

for A, B Borel in $S \times U, [s, t]$ resp. Similar notation will be followed for other $\mathcal{P}(S \times U)$-valued processes. Recall that S being a complete separable metric space, it can be homeomorphically embedded as a dense subset of a *compact* metric space $\bar{S}$. (See Theorem 1.1.1, p. 2 in Borkar, 1995.) As any probability measure on $S \times U$ can be identified with a probability measure on $\bar{S} \times U$ that assigns zero probability to $(\bar{S} - S) \times U$, we may view $\mu(\cdot)$ as a random variable taking values in $\mathcal{U} \stackrel{\text{def}}{=}$ the space of measurable functions $\nu(\cdot) = \nu(\cdot, dy)$ from $[0, \infty)$ to $\mathcal{P}(\bar{S} \times U)$. This space is topologized with the coarsest topology that renders continuous the maps $\nu(\cdot) \in \mathcal{U} \to \int_0^T g(t) \int f d\nu(t) dt \in \mathcal{R}$ for all $f \in C(\bar{S})$, $T > 0$ and $g \in L_2[0,T]$. We shall assume that:

(*) For $f \in C(\bar{S})$, the function

$$(y, z, x) \in S \times U \times \mathcal{R}^d \to \int f(w) p(dw|y, z, x)$$

extends continuously to $\bar{S} \times U \times \mathcal{R}^d$.

Later on we see a specific instance of how this might come to be, viz., in the Euclidean case. With a minor abuse of notation, we retain the original notation f to denote this extension. Finally, we denote by $\mathcal{U}_0 \subset \mathcal{U}$ the subset $\{\mu(\cdot) \in \mathcal{U} : \int_{S \times U} \mu(t, dydz) = 1 \; \forall t\}$ with the relative topology.

Lemma 3. $\mathcal{U}$ *is compact metrizable.*

Proof. For $N \geq 1$, let $\{e_i^N(\cdot), i \geq 1\}$ denote a complete orthonormal basis for $L_2[0, N]$. Let $\{f_j\}$ be countable dense in the unit ball of $C(\bar{S})$. Then it is a convergence determining class for $\mathcal{P}(\bar{S})$ (cf. Appendix C). It can then be easily verified that

$$d(\nu_1(\cdot), \nu_2(\cdot)) \stackrel{\text{def}}{=} \sum_{N \geq 1} \sum_{i \geq 1} \sum_{j \geq 1} 2^{-(N+i+j)} \; \| \int_0^N e_i^N(t) \int f_j d\nu_1(t) dt$$

$$- \int_0^N e_i^N(t) \int f_j d\nu_2(t) dt \| \wedge 1$$

defines a metric on $\mathcal{U}$ consistent with its topology. To show sequential compactness, take $\{\nu_n(\cdot)\} \subset \mathcal{U}$. Recall that $\int f_j d\nu_n(\cdot)|_{[0,N]}, j, n, N \geq 1$, are bounded and therefore relatively sequentially compact in $L_2[0, N]$ endowed with the weak topology. Thus we may use a diagonal argument to pick a subsequence of $\{n\}$, denoted by $\{n\}$ again by abuse of terminology, such that for each j and N, $\int f_j d\nu_n(\cdot)|_{[0,N]} \to \alpha_j(\cdot)|_{[0,N]}$ weakly in $L_2[0, N]$ for some real-valued measurable functions $\{\alpha_j(\cdot), j \geq 1\}$ on $[0, \infty)$ satisfying $\alpha_j(\cdot)|_{[0,N]} \in L_2[0, N]$ $\forall N \geq 1$. Fix j, N. Mimicking the proof of the Banach-Saks theorem (Theorem 1.8.4 of Balakrishnan (1976)), let $n(1) = 1$ and pick $\{n(k)\}$ inductively to satisfy

$$\sum_{j=1}^{\infty} 2^{-j} \max_{1 \leq m < k} | \int_0^N (\int f_j d\nu_{n(k)}(t) - \alpha_j(t))(\int f_j d\nu_{n(m)}(t) - \alpha_j(t)) dt | < \frac{1}{k}.$$

This choice is possible because $\int f_j d\nu_n(\cdot)|_{[0,N]} \to \alpha_j(\cdot)|_{[0,N]}$ weakly in $L_2[0, N]$. Denote by $\| \cdot \|_2$ and $\langle \cdot, \cdot \rangle_2$ the norm and inner product in $L_2[0, N]$. Then for

$j \geq 1$,

$$\|\frac{1}{m}\sum_{k=1}^{m}\int f_j d\nu_{n(k)}(\cdot) - \alpha_j(\cdot)\|_2^2$$

$$\leq \frac{1}{m^2}(2mN^2 + 2\sum_{i=2}^{m}\sum_{\ell=1}^{i-1}|\langle\int f_j d\nu_{n(i)}(\cdot) - \alpha_j(\cdot), \int f_j d\nu_{n(\ell)}(\cdot) - \alpha_j(\cdot)\rangle|)$$

$$\leq \frac{2}{m^2}[mN^2 + 2^j(m-1)] \to 0,$$

as $m \to \infty$. Thus

$$\frac{1}{m}\sum_{k=1}^{m}\int f_j d\nu_{n(k)}(\cdot) \to \alpha_j(\cdot)$$

strongly in $L_2[0, N]$ and hence a.e. along a subsequence $\{m(\ell)\}$ of $\{m\}$. Fix a $t \geq 0$ for which this is true. $\mathcal{P}(\bar{S} \times U)$ is a compact space by Prohorov's theorem – see Appendix C. Let $\nu'(t)$ be a limit point in $\mathcal{P}(\bar{S} \times U)$ of the sequence

$$\{\frac{1}{m(\ell)}\sum_{k=1}^{m(\ell)}\nu_{n(k)}(t), m \geq 1\}.$$

Then $\alpha_j(t) = \int f_j d\nu'(t) \, \forall j$, implying that

$$[\alpha_1(t), \alpha_2(t), \ldots] \in \{[\int f_1 d\nu, \int f_2 d\nu, \ldots] : \nu \in \mathcal{P}(\bar{S} \times U)\}$$

a.e., where the 'a.e.' may be dropped by the choice of a suitable modification of the α_j. By a standard measurable selection theorem (see, e.g., Wagner, 1977), it then follows that there exists a $\nu^*(\cdot) \in \mathcal{U}$ such that $\alpha_j(t) = \int f_j d\nu^*(t) \, \forall t, j$. That is, $d(\nu_n(\cdot), \nu^*(\cdot)) \to 0$. This completes the proof. ∎

We assume as usual the stability condition for $\{x_n\}$: $\sup_n \|x_n\| < \infty$ a.s. In addition, we shall need the following 'stability' condition for $\{Y_n\}$:

(†) Almost surely, for any $t > 0$, the set $\{\mu_s^{s+t}, s \geq 0\}$ remains tight.

Note that while this statement involves both $\{Y_n\}$ and $\{Z_n\}$ via the definition of $\mu(\cdot)$, it is essentially a restriction only on $\{Y_n\}$. This is because $\forall n$, $Z_n \in U$, which is compact. A sufficient condition for (†) when $S = \mathcal{R}^k$ will be discussed later.

Define $\tilde{h}(x, \nu) \stackrel{\text{def}}{=} \int h(x, y)\nu(dy, U)$ for $\nu \in \mathcal{P}(S \times U)$. For $\mu(\cdot)$ as above, consider the non-autonomous o.d.e.

$$\dot{x}(t) = \tilde{h}(x(t), \mu(t)). \tag{6.2.4}$$

Let $x^s(t), t \geq s$, denote the solution to (6.2.4) with $x^s(s) = \bar{x}(s)$, for $s \geq 0$. The following can then be proved along the lines of Lemma 1 of Chapter 2.

Lemma 4. *For any $T > 0$, $\sup_{t \in [s, s+T]} \|\bar{x}(t) - x^s(t)\| \to 0$, a.s.*

We shall also need the following lemma. Let $\mu^n(\cdot) \to \mu^\infty(\cdot)$ in $\mathcal{U}_0$.

Lemma 5. *Let $x^n(\cdot), n = 1, 2, \ldots, \infty$, denote solutions to (6.2.4) corresponding to $\mu(\cdot)$ replaced by $\mu^n(\cdot)$, for $n = 1, 2, \ldots, \infty$. Suppose $x^n(0) \to x^\infty(0)$. Then $\lim_{n \to \infty} \sup_{t \in [t_0, t_0 + T]} \|x^n(t) - x^\infty(t)\| \to 0$ for every $t_0, T > 0$.*

Proof. Take $t_0 = 0$ for simplicity. By our choice of the topology for $\mathcal{U}_0$,

$$\int_0^t g(t) \int f d\mu^n(s) ds - \int_0^t g(t) \int f d\mu^\infty(s) ds \to 0$$

for bounded continuous $g : [0, t] \to \mathcal{R}$, $f : \bar{S} \to \mathcal{R}$. Hence

$$\int_0^t \int \tilde{f}(s, \cdot) d\mu^n(s) ds - \int_0^t \int \tilde{f}(s, \cdot) d\mu^\infty(s) ds \to 0$$

for all bounded continuous $\tilde{f} : [0, t] \times \bar{S} \to \mathcal{R}$ of the form

$$\tilde{f}(s, w) = \sum_{m=1}^N a_m g_m(s) f_m(w)$$

for some $N \geq 1$, scalars a_i and bounded continuous real-valued functions g_i, f_i on $[0, t], \bar{S}$ resp., for $1 \leq i \leq N$. By the Stone-Weierstrass theorem, such functions can uniformly approximate any $\tilde{f} \in C([0, T] \times \bar{S})$. Thus the above convergence holds true for all such $\tilde{f}$, implying that $t^{-1} d\mu^n(s) ds \to t^{-1} d\mu^\infty(s) ds$ in $\mathcal{P}(\bar{S} \times [0, t])$ and hence in $\mathcal{P}(S \times [0, t])$. Thus in particular

$$\left\| \int_0^t (\tilde{h}(x^\infty(s), \mu^n(s)) - \tilde{h}(x^\infty(s), \mu^\infty(s))) ds \right\| \to 0.$$

As a function of t, the integral on the left is equicontinuous and pointwise bounded. By the Arzela-Ascoli theorem, this convergence must in fact be uniform for t in a compact set. Now for $t > 0$,

$$\|x^n(t) - x^\infty(t)\| \leq \|x^n(0) - x^\infty(0)\|$$
$$+ \int_0^t \|\tilde{h}(x^n(s), \mu^n(s)) - \tilde{h}(x^\infty(s), \mu^n(s))\| ds$$
$$+ \left\| \int_0^t (\tilde{h}(x^\infty(s), \mu^n(s)) - \tilde{h}(x^\infty(s), \mu^\infty(s))) ds \right\|$$
$$\leq \|x^n(0) - x^\infty(0)\| + L \int_0^t \|x^n(s) - x^\infty(s)\| ds$$
$$+ \left\| \int_0^t (\tilde{h}(x^\infty(s), \mu^n(s)) - \tilde{h}(x^\infty(s), \mu^\infty(s))) ds \right\|.$$

By the Gronwall inequality, there exists $K_T > 0$ such that

$$\sup_{t \in [0,T]} \|x^n(t) - x^\infty(t)\|$$

$$\leq K_T \Big(\|x^n(0) - x^\infty(0)\|$$

$$+ \sup_{t \in [0,T]} \| \int_0^t (\tilde{h}(x^\infty(s), \mu^n(s)) - \tilde{h}(x^\infty(s), \mu^\infty(s)))ds\| \Big).$$

In view of the foregoing, this leads to the desired conclusion. ∎

6.3 Averaging the natural timescale: main results

The key consequence of (†) that we require is the following:

Lemma 6. *Almost surely, every limit point of $(\mu_s^{s+t}, x^s(\cdot))$ for $t > 0$ as $s \to \infty$ is of the form $(\tilde{\mu}_0^t, \tilde{x}(\cdot))$, where*

- $\tilde{\mu}(\cdot)$ *satisfies* $\tilde{\mu}(t) \in D(\tilde{x}(t))$, *and*
- $\tilde{x}(\cdot)$ *satisfies (6.2.4) with $\mu(\cdot)$ replaced by $\tilde{\mu}(\cdot)$.*

Proof. Let $\{f_i\}$ be a countable set of bounded continuous functions $S \to R$ that is a convergence determining class for $\mathcal{P}(S)$. By replacing each f_i by $a_i f_i + b_i$ for suitable $a_i, b_i > 0$, we may suppose that $0 \leq f_i(\cdot) \leq 1$ for all i. For each i,

$$\xi_n^i \overset{\text{def}}{=} \sum_{m=1}^{n-1} a(m)(f_i(Y_{m+1}) - \int f_i(w)p(dw|Y_m, Z_m, x_m)),$$

is a zero mean martingale with $\sup_n E[\|\xi_n^i\|^2] \leq \sum_n a(n)^2 < \infty$. By the martingale convergence theorem (cf. Appendix C), it converges a.s. Let $\tau(n, s) \overset{\text{def}}{=} \min\{m \geq n : t(m) \geq t(n) + s\}$ for $s \geq 0, n \geq 0$. Then as $n \to \infty$,

$$\sum_{m=n}^{\tau(n,t)} a(m)(f_i(Y_{m+1}) - \int f_i(w)p(dw|Y_m, Z_m, x_m)) \to 0, \quad \text{a.s.}$$

for $t > 0$. By our choice of $\{f_i\}$ and the fact that $\{a(n)\}$ are eventually nonincreasing (this is the only time the latter property is used),

$$\sum_{m=n}^{\tau(n,t)} (a(m) - a(m+1))f_i(Y_{m+1}) \to 0, \quad \text{a.s.}$$

Thus

$$\sum_{m=n}^{\tau(n,t)} a(m)(f_i(Y_m) - \int f_i(w)p(dw|Y_m, Z_m, x_m)) \to 0, \quad \text{a.s.}$$

Dividing by $\sum_{m=n}^{\tau(n,t)} a(m) \geq t$ and using (*) and the uniform continuity of $p(dw|y, z, x)$ in x on compacts, we obtain

$$\int_{t(n)}^{t(n)+t} \int (f_i(y) - \int f_i(w)p(dw|y, z, \bar{x}(s))\mu(s, dydz))ds \to 0, \text{ a.s.}$$

Fix a sample point in the probability one set on which the convergence above holds for all i. Let $(\tilde{\mu}(\cdot), \tilde{x}(\cdot))$ be a limit point of $(\mu(s + \cdot), x^s(\cdot))$ in $\mathcal{U} \times C([0, \infty); \mathcal{R}^d)$ as $s \to \infty$. Then the convergence above leads to

$$\int_0^t \int (f_i(y) - \int f_i(w)p(dw|y, z, \tilde{x}(s)))\tilde{\mu}(s, dydz)ds = 0 \; \forall i. \qquad (6.3.1)$$

By (†), $\tilde{\mu}_0^t(S \times U \times [0, t]) = 1 \; \forall t$ and thus it follows that $\tilde{\mu}_s^t(S \times U \times [s, t]) = 1 \; \forall t > s \geq 0$. By Lebesgue's theorem (see Appendix A), one then has $\tilde{\mu}(t)(S \times U) = 1$ for a.e. t. A similar application of Lebesgue's theorem in conjunction with (6.3.1) shows that

$$\int (f_i(y) - \int f_i(w)p(dw|y, z, \tilde{x}(t)))\tilde{\mu}(t, dydz) = 0 \; \forall i,$$

for a.e. t. The qualification 'a.e. t' here may be dropped throughout by choosing a suitable modification of $\tilde{\mu}(\cdot)$. By our choice of $\{f_j\}$, this leads to

$$\tilde{\mu}(t, dw \times U) = \int p(dw|y, z, \tilde{x}(t))\tilde{\mu}(t, dydz).$$

The claim follows from this and Lemma 5.   ∎

Combining Lemmas 3 – 6 immediately leads to our main result:

Theorem 7. *Almost surely, $\{\bar{x}(s + \cdot), s \geq 0\}$ converge to an internally chain transitive invariant set of the differential inclusion*

$$\dot{x}(t) \in \hat{h}(x(t)), \qquad (6.3.2)$$

as $s \to \infty$, where $\hat{h}(x) \stackrel{def}{=} \{\tilde{h}(x, \nu) : \nu \in D(x)\}$. In particular $\{x_n\}$ converge a.s. to such a set.

For special cases, more can be said, e.g., in the following:

Corollary 8. *Suppose there is no additional control process $\{Z_n\}$ in (6.2.2) and for each $x \in \mathcal{R}^d$ and $x_n \equiv x \; \forall n$, $\{Y_n\}$ is an ergodic Markov process with a unique invariant probability measure $\nu(x) = \nu(x, dy)$. Then (6.3.2) above may be replaced by the o.d.e.*

$$\dot{x}(t) = \tilde{h}(x(t), \nu(x(t))). \qquad (6.3.3)$$

If $p(dw|y,x)$ denotes the transition kernel of this ergodic Markov process, then $\nu(x)$ is characterized by

$$\int \left(f(y) - \int f(w)p(dw|y,x) \right) \nu(x,dy) = 0$$

for bounded $f \in C(S)$. Since this equation is preserved under convergence in $\mathcal{P}(S)$, it follows that $x \to \nu(x)$ is a continuous map. This guarantees the existence of solutions to (6.3.3) by standard o.d.e. theory, though not their uniqueness. In general, the solution set for a fixed initial condition will be a nonempty compact subset of $C([0,\infty);\mathcal{R}^d)$. For uniqueness, we need $\tilde{h}(\cdot,\nu(\cdot))$ to be Lipschitz, which requires additional information about ν and the transition kernel p.

Many of the developments of the previous chapters have their natural counterparts for (6.2.1). For example, the first stability criterion of Chapter 3 (Theorem 7) has the following natural extension, stated here for the simpler case when assumptions of Corollary 8 hold. The notation is as above.

Theorem 9. *Suppose the limit*

$$\hat{h}(x) \stackrel{def}{=} \lim_{a \uparrow \infty} \frac{\tilde{h}(\frac{x(t)}{a}, \nu(\frac{x(t)}{a}))}{a}$$

exists uniformly on compacts, and furthermore, the o.d.e.

$$\dot{x}(t) = \hat{h}(x(t))$$

is well posed and has the origin as the unique globally asymptotically stable equilibrium. Then $\sup_n \|x_n\| < \infty$ *a.s.*

As an interesting 'extension', suppose $\{Y_n, -\infty < n < \infty\}$ is a not necessarily Markov process, with the conditional law of Y_n given by

$$Y^{-n} \stackrel{def}{=} [Y_n, Y_{n-1}, Y_{n-2}, \ldots]$$

being a continuous map $S^\infty \to \mathcal{P}(S)$ independent of n. Then $\{Y^{-n}\}$ is a time-homogeneous Markov process. Let $\gamma : S^\infty \to S$ denote the map that takes $[s_1, s_2, \ldots] \in S^\infty$ to $s_1 \in S$. Replacing S by S^∞, $\{Y_n\}$ by $\{Y^{-n}\}$, and $h(x, \cdot)$ by $h(x, \gamma(\cdot))$, we can reduce this case to the one studied above. The results above then apply as long as the technical assumptions made from time to time can be verified. The case of stationary $\{Y_n\}$ (or something that is nearly the same, viz., the case when the appropriate time averages exist) is in fact the most extensively studied case in the literature (see, e.g., Kushner and Yin, 2003).

6.4 Concluding remarks

We conclude this chapter with a sufficient condition for (†) when $S = \mathcal{R}^m$ for some $m \geq 1$. The condition is that there exists a $V \in C(\mathcal{R}^m)$ such that $\lim_{\|x\| \to \infty} V(x) = \infty$ and furthermore,

$$\sup_n E[V(Y_n)^2] < \infty, \qquad (6.4.1)$$

and for some compact $B \subset \mathcal{R}^m$ and scalar $\epsilon_0 > 0$,

$$E[V(Y_{n+1})|\mathcal{F}_n] \leq V(Y_n) - \epsilon_0, \qquad (6.4.2)$$

a.s. on $\{Y_n \notin B\}$.

In the framework of sections 6.2 and 6.3, we now replace S by $\mathcal{R}^m$ and $\bar{S}$ by $\bar{\mathcal{R}}^m \stackrel{\text{def}}{=}$ the one-point compactification of $\bar{\mathcal{R}}^m$ with the additional 'point at infinity' denoted simply by '∞'. We assume that $p(dw|y, z, x) \to \delta_\infty$ in $\mathcal{P}(\mathcal{R}^m)$ as $\|y\| \to \infty$ uniformly in x, z.

Lemma 10. *Any limit point* $(\mu^*(\cdot), x^*(\cdot))$ *of* $(\mu(s + \cdot), \bar{x}(s + \cdot))$ *as* $s \to \infty$ *in* $\mathcal{U} \times C([0, \infty); \mathcal{R}^d)$ *is of the form*

$$\mu^*(t) = a(t)\tilde{\mu}(t) + (1 - a(t))\delta_\infty, \ t \geq 0,$$

where $a(\cdot)$ *is a measurable function* $[0, \infty) \to [0, 1]$ *and* $\tilde{\mu}(t) \in D(x^*(t)) \ \forall t$.

Proof. Let $\{f_i\}$ denote a countable convergence determining class of functions for $\mathcal{R}^m$ satisfying $\lim_{\|x\| \to \infty} |f_i(x)| = 0$ for all i. Thus they extend continuously to $\bar{\mathcal{R}}^m$ with value zero at ∞. Also, note that by our assumption above, $\lim_{\|y\| \to \infty} \int f_i(w)p(dw|y, z, x) \to 0$ uniformly in z, x, which verifies (*). Argue as in the proof of Lemma 6 to conclude that

$$\int (f_i(y) - \int f_i(w)p(dw|y, z, x^*(t)))\mu^*(t, dydz) = 0 \ \forall i,$$

for all t, a.s. Write $\mu^*(t) = a(t)\tilde{\mu}(t) + (1 - a(t))\delta_\infty$ with $a(\cdot) : [0, \infty) \to [0, 1]$ a measurable map. This is always possible (the decomposition being in fact unique for those t for which $a(t) > 0$). Then when $a(t) > 0$, the above reduces to

$$\int (f_i(y) - \int f_i(w)p(dw|y, z, x^*(t)))\tilde{\mu}(t, dydz) = 0 \ \forall i,$$

for all t. Thus $\tilde{\mu}(t) \in D(x^*(t))$ when $a(t) > 0$. When $a(t) = 0$, the choice of $\tilde{\mu}(t)$ is arbitrary and it may be chosen so that it is in $D(x^*(t))$. The claim follows. ∎

Corollary 11. *Condition (†) holds. That is, almost surely, for any* $t > 0$, *the set* $\{\mu_s^{s+t}, \ s \geq 0\}$ *remains tight.*

Proof. Replacing f_i by V in the proof of Lemma 6 and using (6.4.1) to justify the use of the martingale convergence theorem therein, we have

$$\lim_{s \to \infty} \int_0^t \int \int (V(w)p(dw|y, z, \bar{x}(s+r)) - V(y))\mu(s+r, dydz)dr = 0,$$

a.s. Fix a sample point where this and Lemma 10 hold. Extend the map

$$\psi : (x, y, z) \in \mathcal{R}^d \times \bar{\mathcal{R}}^m \times U \to \int V(w)p(dw|y, z, x) - V(y)$$

to $\mathcal{R}^d \times \bar{\mathcal{R}}^m \times U$ by setting $\psi(x, \infty, z) = -\epsilon_0$, whence it is upper semicontinuous. Thus taking the above limit along an appropriate subsequence along which $(\mu(s + \cdot), \bar{x}(s + \cdot)) \to (\mu^*(\cdot), x^*(\cdot))$ (say), we get

$$
\begin{aligned}
0 \le\ & -\epsilon_0 \int_0^t (1 - a(s))ds \\
& + \int_0^t a(s) \left(\int \int (V(w)p(dw|y, z, x^*(s)) - V(y))\mu^*(s, dydz) \right) ds \\
=\ & -\epsilon_0 \int_0^t (1 - a(s))ds,
\end{aligned}
$$

by Lemma 10. Thus $a(s) = 1$ a.e., where the 'a.e.' may be dropped by taking a suitable modification of $\mu^*(\cdot)$. This implies that the convergence of $\mu(s(n) + \cdot)$ to $\mu^*(\cdot)$ is in fact in $\mathcal{U}_0$. This establishes (†). ∎

7

Asynchronous Schemes

7.1 Introduction

Until now we have been considering the case where all components of x_n are updated simultaneously at time n and the outcome is immediately available for the next iteration. There may, however, be situations when different components are updated by possibly different processors (these could be in different locations, e.g., in remote sensing applications). Furthermore, each of these components may be running on its own 'clock' and exchanging information with the others with some communication delays. This is the distributed, asynchronous implementation of the algorithm. The theory we have developed so far does not apply automatically any more and some work is needed to figure out when it does and when it doesn't. Another important class of problems which lands us into a similar predicament consists of the multiagent learning or optimization schemes when each component actually corresponds to a different autonomous agent and the aforementioned complications arise naturally. Yet another situation involves the 'on-line' algorithms for control or estimation of a Markov chain in which we have a one-to-one correspondence between the components of x_n and the state space of the chain (i.e., the ith component of x_n is a quantity associated with state i of the chain), and the ith component gets updated only when state i is visited. We shall see examples of this later on.

A mathematical model that captures the aspects above is as follows: Letting $x_n = [x_n(1), \ldots, x_n(d)]$, the ith (for $1 \leq i \leq d$) component is updated in our original scheme according to

$$x_{n+1}(i) = x_n(i) + a(n)[h_i(x_n) + M_{n+1}(i)], \quad n \geq 1, \qquad (7.1.1)$$

where $h_i, M_n(i)$ are the ith components of h, M_n resp., for $n \geq 1$. We replace this by

$$x_{n+1}(i) = x_n(i) + a(\nu(i,n))I\{i \in Y_n\}$$
$$\times [h_i(x_{n-\tau_{1i}(n)}(1), \ldots, x_{n-\tau_{di}(n)}(d)) + M_{n+1}(i)], \quad (7.1.2)$$

for $n \geq 0$. Here:

(i) Y_n is a random subset of the index set $\{1, \ldots, d\}$, indicating the subset of components which are updated at time n,

(ii) $0 \leq \tau_{ij}(n) \leq n$ is the delay faced by 'processor' j in receiving the output of processor i at time n. In other words, at time n, processor j knows $x_{n-\tau_{ij}(n)}(i)$, but not $x_m(i)$ for $m > n - \tau_{ij}(n)$ (or does know some of them but does not realize they are more recent!).

(iii) $\nu(i,n) \stackrel{\text{def}}{=} \sum_{m=0}^n I\{i \in Y_m\}$, i.e., the number of times the ith component was updated up until time n.

Note that the ith processor needs to know only its *local clock* $\{\nu(i,n)\}$ and not the *global clock* $\{n\}$. In fact the global clock can be a complete artifice as long as causal relationships are respected. One usually has

$$\liminf_{n \to \infty} \frac{\nu(i,n)}{n} > 0. \quad (7.1.3)$$

This means that all components are being updated *comparably often*. A simple sufficient condition for (7.1.3) would be that $\{Y_n\}$ is an irreducible and hence positive recurrent Markov chain on the power set of $\{1, \ldots, d\}$. (More generally, it could be a *controlled* Markov chain on this state space with the property that any stationary policy leads to an irreducible chain with a stationary distribution that assigns a probability $\geq \delta$ to each state, for some $\delta > 0$. More on this later.) Note in particular that this condition ensures that $\nu(i,n) \uparrow \infty$ for all i, i.e., each component is updated infinitely often. For the purposes of the next section, this is all we need.

Define

$$\mathcal{F}_n = \sigma(x_m, M_m, Y_m, \tau_{ij}(m), 1 \leq i, j \leq d, m \leq n), \ n \geq 0.$$

We assume that

$$E[M_{n+1}(i)|\mathcal{F}_n] = 0,$$
$$E[|M_{n+1}(i)|^2|\mathcal{F}_n] \leq K(1 + \sup_{m \leq n} ||x_m||^2), \quad (7.1.4)$$

where $1 \leq i \leq d$, $n \geq 0$, and $K > 0$ is a suitable constant.

Usually it makes sense to assume $\tau_{ii}(m) = 0$ for all i and $m \leq n$, and we shall do so (implying that a processor has its own past outputs immediately available). This, however, is not essential for the analysis that follows.

The main result here is that under suitable conditions, the interpolated iterates track a time-dependent o.d.e. of the form

$$\dot{x}(t) = \Lambda(t)h(x(t)), \tag{7.1.5}$$

where $\Lambda(\cdot)$ is a matrix-valued measurable process such that $\Lambda(t)$ for each t is a diagonal matrix with nonnegative diagonal entries. These in some sense reflect the relative 'instantaneous' rates with which the different components get updated. Our treatment follows Borkar (1998). See also Kushner and Yin (1987a, 1987b).

7.2 Asymptotic behavior

As usual, we shall start by assuming

$$\sup_n \|x_n\| < \infty, \text{ a.s.} \tag{7.2.1}$$

We shall also simplify the situation by assuming that there are no delays, i.e., $\tau_{ij}(n) \equiv 0 \ \forall i, j, n$. The effect of delays will be considered separately later on. Thus (7.1.2) becomes

$$x_{n+1}(i) = x_n(i) + a(\nu(i,n))I\{i \in Y_n\}$$
$$\times [h_i(x_n(1), \ldots, x_n(d)) + M_{n+1}(i)], \tag{7.2.2}$$

for $n \geq 0$. Let $\bar{a}(n) \overset{\text{def}}{=} \max_{i \in Y_n} a(\nu(i,n)) > 0, n \geq 0$. Then it is easy to verify that $\sum_n \bar{a}(n) = \infty, \sum_n \bar{a}(n)^2 < \infty$ a.s.: we have, for any fixed $i, 1 \leq i \leq d$,

$$\sum_n \bar{a}(n) \geq \sum_n a(\nu(i,n))I\{i \in Y_n\}$$
$$= \sum_n a(n) = \infty, \text{ and}$$
$$\sum_n \bar{a}(n)^2 \leq \sum_n \sum_i a(\nu(i,n))^2 I\{i \in Y_n\}$$
$$\leq d \sum_n a(n)^2 < \infty. \tag{7.2.3}$$

This implies in particular that $\{\bar{a}(n)\}$ is a legitimate stepsize schedule, albeit random. (See the comments following Theorem 2 of Chapter 2.) Rewrite (7.2.2) as

$$x_{n+1}(i) = x_n(i) + \bar{a}(n)q(i,n)$$
$$\times [h_i(x_n(1), \ldots, x_n(d)) + M_{n+1}(i)], \tag{7.2.4}$$

where $q(i,n) \overset{\text{def}}{=} (a(\nu(i,n))/\bar{a}(n))I\{i \in Y_n\} \in (0,1] \ \forall n$. As before, define $t(0) = 0, t(n) = \sum_{m=0}^n \bar{a}(m), n \geq 1$. Define $\bar{x}(t), t \geq 0$, by $\bar{x}(t(n)) = x_n, n \geq 0$, with linear interpolation on each interval $I_n \stackrel{\text{def}}{=} [t(n), t(n+1)]$. For $1 \le i \le d$, define $u_i(t), t \ge 0$, by $u_i(t) = q(i, n)$ for $t \in [t(n), t(n+1))$, $n \ge 0$. Let $\lambda(t) = \text{diag}(u_1(t), \ldots, u_d(t)), t \ge 0$, and $x^s(t), t \ge s$, the unique solution to the non-autonomous o.d.e.

$$\dot{x}^s(t) = \lambda(t)h(x^s(t)), \quad t \ge s.$$

The following lemma then holds by familiar arguments.

Lemma 1. *For any $T > 0$,*

$$\lim_{s \to \infty} \sup_{t \in [s, s+T]} \|\bar{x}(t) - x^s(t)\| = 0, \quad a.s.$$

This immediately leads to:

Theorem 2. *Almost surely, any limit point of $\bar{x}(s + \cdot)$ in $C([0, \infty); \mathcal{R}^d)$ as $s \uparrow \infty$ is a solution of a non-autonomous o.d.e.*

$$\dot{x}(t) = \Lambda(t)h(x(t)), \tag{7.2.5}$$

where $\Lambda(\cdot)$ is a $d \times d$-dimensional diagonal matrix-valued measurable function with entries in $[0, 1]$ on the diagonal.

Proof. View $u(\cdot) \stackrel{\text{def}}{=} [u_1(\cdot), \ldots, u_d(\cdot)]$ as an element of $\mathcal{V} \stackrel{\text{def}}{=}$ the space of measurable maps $y(\cdot) : [0, \infty) \to [0, 1]^d$ with the coarsest topology that renders continuous the maps

$$y(\cdot) \to \int_0^t \langle g(s), y(s) \rangle ds,$$

for all $t > 0$, $g(\cdot) \in L_2([0, t]; \mathcal{R}^d)$. A standard application of the Banach–Alaoglu theorem (see Appendix A) shows that this is a compact space, metrizable by the metric

$$\rho(y_1(\cdot), y_2(\cdot)) \stackrel{\text{def}}{=} \sum_{n=1}^{\infty} \sum_{m=1}^{\infty} 2^{-(n+m)}$$
$$\times \min\left(1, \left| \int_0^n \langle y_1(t), e_m^n(t) \rangle dt - \int_0^n \langle y_2(t), e_m^n(t) \rangle dt \right| \right),$$

where $\{e_m^n(\cdot), m \ge 1\}$ is a complete orthonormal basis for $L_2([0, n]; \mathcal{R}^d)$. Relative compactness of $\bar{x}(t + \cdot), t \ge 0$, in $C([0, \infty); \mathcal{R}^d)$ is established as before. Consider $t_n \to \infty$ such that $\bar{x}(t_n + \cdot) \to x^*(\cdot)$ (say) in $C([0, \infty); \mathcal{R}^d)$. By dropping to a subsequence if necessary, assume that $u(t_n + \cdot) \to u^*(\cdot)$ in $\mathcal{V}$. Let $\Lambda(\cdot)$ denote the diagonal matrix with ith diagonal entry $= u^*(i)$. By Lemma 1,

$$\bar{x}(t_n + s) - \bar{x}(t_n + r) = \int_r^s \lambda(t_n + z)h(\bar{x}(t_n + z))dz + o(1), \quad s > r \ge 0.$$

Letting $n \to \infty$ in this equation, familiar arguments from Chapter 6 yield

$$x^*(s) - x^*(r) = \int_r^s \Lambda(z)h(x^*(z))dz, \; s > r \geq 0.$$

This completes the proof.   ∎

7.3 Effect of delays

Next we shall consider the effect of delays. Specifically, we look for conditions under which Theorem 2 will cóntinue to hold for $\{x_n\}$ given by (7.1.2) instead of (7.2.2). We shall assume that each output $x_n(j)$ of the jth processor is transmitted to the ith processor for any pair (i,j) almost surely, though we allow for some outputs to be 'lost' in transit. The situation where not all outputs are transmitted can also be accommodated by equating unsent outputs with lost ones. In this case our requirement boils down to infinitely many outputs being transmitted. At the receiver end, we assume that the ith processor receives infinitely many outputs sent by j almost surely, though not necessarily in the order sent. This leaves two possibilities at the receiver: Either the messages are 'time-stamped' and the receiver can re-order them and use at each iteration the one sent most recently, or they are not and the receiver uses the one received most recently. Our analysis allows for both possibilities, subject to the additional condition that $a(m+n) \leq \kappa a(n)$ for all $m, n \geq 0$ and some $\kappa > 0$. This is a very mild restriction.

Comparing (7.1.2) with (7.2.2), one notes that the delays introduce in the $(n+1)$st iteration of the ith component an additional error of

$$a(\nu(i,n))I\{i \in Y_n\}$$
$$\times (h_i(x_{n-\tau_{1i}(n)}(1), \cdots, x_{n-\tau_{di}(n)}(d)) - h_i(x_n(1), \cdots, x_n(d))).$$

Our aim will be to find conditions under which this error is $o(a(n))$. If so, one can argue as in the extension at the start of section 2.2 and conclude that Theorem 2 continues to hold with (7.1.2) in place of (7.2.2). Since $h(\cdot)$ is Lipschitz, the above error is bounded by a constant times

$$a(\nu(i,n)) \sum_j |x_n(j) - x_{n-\tau_{ji}(n)}(j)|.$$

We shall consider each summand (say, the jth) separately. This is bounded by

$$|\sum_{m=n-\tau_{ji}(n)}^{n-1} a(\nu(j,m))I\{j \in Y_m\}h_j(x_{m-\tau_{1j}(m)}(1), \cdots, x_{m-\tau_{dj}(m)}(d))|$$

$$+|\sum_{m=n-\tau_{ji}(n)}^{n-1} a(\nu(j,m))I\{j \in Y_m\}M_{m+1}(j)|. \tag{7.3.1}$$

We shall impose the mild assumption:

$$n - \tau_{k\ell}(n) \uparrow \infty \text{ a.s.} \tag{7.3.2}$$

for all k, ℓ. As before, (7.1.4), (7.2.1) and (7.2.3) together imply that the sum $\sum_{m=0}^{n} a(\nu(j,m)) I\{j \in Y_m\} M_{m+1}(j)$ converges a.s. for all j. In view of (7.3.2), we then have

$$\left| \sum_{m=n-\tau_{ji}(n)}^{n-1} a(\nu(j,m)) I\{j \in Y_m\} M_{m+1}(j) \right| = o(1),$$

implying that

$$a(\nu(i,n)) \left| \sum_{m=n-\tau_{ji}(n)}^{n-1} a(\nu(j,m)) I\{j \in Y_m\} M_{m+1}(j) \right| = o(a(\nu(i,n))).$$

Under (7.2.1), the first term of (7.3.1) can be almost surely bounded from above by a (sample path dependent) constant times

$$\sum_{m=n-\tau_{ji}(n)}^{n-1} a(\nu(j,m)).$$

(See, e.g., Chapter 2.) Since $a(n) \le \kappa a(m)$ for $m \le n$, this in turn is bounded by $\kappa a(\nu(j, n - \tau_{ji}(n))) \tau_{ji}(n)$ for large n. Thus we are done if this quantity is $o(1)$. Note that by (7.3.2), this is certainly so if the delays are bounded. More generally, suppose that

$$\frac{\tau_{ji}(n)}{n} \to 0 \text{ a.s.}$$

This is a perfectly reasonable condition and can be recast as

$$\frac{n - \tau_{ji}(n)}{n} \to 1 \text{ a.s.} \tag{7.3.3}$$

Note that this implies (7.3.2). We further assume that

$$\limsup_{n \to \infty} \sup_{y \in [x,1]} \frac{a(\lfloor yn \rfloor)}{a(n)} < \infty \; \forall i, \tag{7.3.4}$$

for $0 < x \le 1$. This is also quite reasonable as it is seen to hold for most standard examples of $\{a(n)\}$. Furthermore, this implies that whenever (7.1.3) and (7.3.3) hold,

$$\limsup_{n \to \infty} \frac{a(\nu(j, n - \tau_{k\ell}(n)))}{a(n)} < \infty$$

for all k, ℓ. Thus our task reduces to showing $a(n) \tau_{k\ell}(n) = o(1)$ for all k, ℓ. Assume the following:

(†) There exists $\eta > 0$ and a nonnegative integer valued random variable $\bar{\tau}$ such that:

- $a(n) = o(n^{-\eta})$ and
- $\bar{\tau}$ stochastically dominates all $\tau_{k\ell}(n)$ and satisfies

$$E[\bar{\tau}^{\frac{1}{\eta}}] < \infty.$$

All standard examples of $\{a(n)\}$ satisfy the first condition with a natural choice of η, e.g., for $a(n) = n^{-1}$, take $\eta = 1 - \epsilon$ for any $\epsilon \in (0, 1)$. The second condition is easily verified, e.g., if the tails of the delay distributions show uniform exponential decay. Under (†),

$$
\begin{aligned}
P(\tau_{k\ell}(n) \geq n^{\eta}) &\leq P(\bar{\tau} \geq n^{\eta}) \\
&= P(\bar{\tau}^{\frac{1}{\eta}} \geq n),
\end{aligned}
$$

leading to

$$
\begin{aligned}
\sum_n P(\tau_{k\ell}(n) \geq n^{\eta}) &\leq \sum_n P(\bar{\tau}^{\frac{1}{\eta}} \geq n) \\
&= E[\bar{\tau}^{\frac{1}{\eta}}] \\
&< \infty.
\end{aligned}
$$

By the Borel–Cantelli lemma, one then has

$$P(\tau_{k\ell}(n) \geq n^{\eta} \text{ i.o.}) = 0.$$

Coupled with the first part of (†), this implies $a(n)\tau_{k\ell}(n) = o(1)$. We have proved:

Theorem 3. *Under assumptions (7.3.3), (7.3.4) and (†), the conclusions of Theorem 2 also hold when $\{x_n\}$ are generated by (7.1.2).*

The following discussion provides some intuition as to why the delays are 'asymptotically negligible' as long as they are not 'arbitrarily large', in the sense of (7.3.3). Recall our definition of $\{t(n)\}$. Note that the passage from the original discrete time count $\{n\}$ to $\{t(n)\}$ implies a time scaling. In fact this is a 'compression' of the time axis because the successive differences $t(n+1) - t(n)$ tend to zero. An interval $[n, n+1, \ldots, n+N]$ on the original time axis gets mapped to $[t(n), t(n+1), \ldots, t(n+N)]$ under this scaling. As $n \to \infty$, the width of the former remains constant at N, whereas that of the latter, $t(n+N) - t(n)$, tends to zero. That is, intervals of a fixed length get 'squeezed out' in the limit as $n \to \infty$. Since the approximating o.d.e. we are looking at is operating on the transformed timescale, the net variation of its trajectories over these intervals is less and less as $n \to \infty$, hence so is the case of interpolated iterates $\bar{x}(\cdot)$. In other words, the error between the most recent iterate from a processor and one received with a bounded delay is asymptotically negligible. The same intuition carries over for possibly unbounded delays that satisfy (7.3.3).

7.4 Convergence

We now consider several instances where Theorems 2 and 3 can be strengthened.

(i) The first and perhaps the most important case is when convergence is obtained just by a judicious choice of stepsize schedule. Let $\Lambda(t)$ above be written as $\mathrm{diag}(\eta_1(t), \ldots, \eta_d(t))$, i.e., the diagonal matrix with the ith diagonal entry equal to $\eta_i(t)$. For $n \geq 0, s > 0$, let

$$N(n, s) \stackrel{\text{def}}{=} \min\{m > n : t(m) \geq t(n) + s\} > n.$$

From the manner in which $\Lambda(\cdot)$ was obtained, it is clear that there exist $\{n(k)\} \subset \{n\}$ such that

$$\int_t^{t+s} \eta_i(y)dy = \lim_{k \to \infty} \sum_{m=n(k)}^{N(n(k),s)} \frac{a(\nu(i,m))I\{i \in Y_m\}}{\bar{a}(m)}\bar{a}(m)$$

$$= \lim_{k \to \infty} \sum_{m=\nu(i,n(k))}^{\nu(i,N(n(k),s))} a(m) \quad \forall i.$$

Thus

$$\frac{\int_t^{t+s} \eta_i(y)dy}{\int_t^{t+s} \eta_j(y)dy} = \lim_{k \to \infty} \frac{\sum_{m=\nu(i,n(k))}^{\nu(i,N(n(k),s))} a(m)}{\sum_{m=\nu(j,n(k))}^{\nu(j,N(n(k),s))} a(m)} \quad \forall i, j. \tag{7.4.1}$$

Suppose we establish that under (7.1.3), the right-hand side of (7.4.1) is always 1. Then (7.2.5) is of the form

$$\dot{x}(t) = \alpha h(x(t)),$$

for a scalar $\alpha > 0$. This is simply a time-scaled version of the o.d.e.

$$\dot{x}(t) = h(x(t)) \tag{7.4.2}$$

and hence has exactly the same trajectories. Thus the results of Chapter 2 apply. See Borkar (1998) for one such situation.

(ii) The second important situation is when the o.d.e. (7.1.5) has the same asymptotic behaviour as (7.4.2) purely because of the specific structure of $h(\cdot)$. Consider the following special case of the scenario of Corollary 3 of Chapter 2, with a continuously differentiable Liapunov function $V(\cdot)$ satisfying

$$\lim_{||x||\to\infty} V(x) = \infty, \ \langle h(x), \nabla V(x) \rangle < 0 \ \text{ whenever } \ h(x) \neq 0.$$

Suppose

$$\liminf_{t\to\infty} \eta_i(t) \geq \epsilon > 0 \ \forall i, \qquad (7.4.3)$$

and

$$\langle h(x), \Gamma \nabla V(x) \rangle < 0 \ \text{ whenever } \ h(x) \neq 0,$$

for all $d \times d$ diagonal matrices Γ satisfying $\Gamma \geq \epsilon I_d$, I_d being the $d \times d$ identity matrix. By (7.4.3), $\Lambda(t) \geq \epsilon I_d \ \forall t$. Then exactly the same argument as for Corollary 3 of Chapter 2 applies to (7.1.5), leading to the conclusion that $x_n \to \{x : h(x) = 0\}$ a.s. An important instance of this lucky situation is the case when $h(x) = -\nabla F(x)$ for some $F(\cdot)$, whence for $V(\cdot) \equiv F(\cdot)$ and Γ as above,

$$\langle h(x), \Gamma \nabla F(x) \rangle \leq -\epsilon ||\nabla F(x)||^2 < 0$$

outside $\{x : \nabla F(x) = 0\}$.

Another example is the case when $h(x) = F(x) - x$ for some $F(\cdot)$ satisfying

$$||F(x) - F(y)||_\infty \leq \beta ||x - y||_\infty \ \forall x, y, \qquad (7.4.4)$$

with $||x||_\infty \overset{\text{def}}{=} \max_i |x_i|$ for $x = [x_1, \ldots, x_d]$ and $\beta \in (0,1)$. That is, F is a *contraction w.r.t. the max-norm*. In this case, it is known from the contraction mapping theorem that there is a unique x^* such that $F(x^*) = x^*$, i.e., a unique equilibrium point for (7.4.2). Furthermore, a direct calculation shows that $V(x) \overset{\text{def}}{=} ||x - x^*||_\infty$ serves as a Liapunov function, albeit a non-smooth one. In fact,

$$||x(t) - x^*||_\infty \downarrow 0. \qquad (7.4.5)$$

See Theorem 2 of Chapter 10 for details. Now,

$$\Gamma(F(x) - x) = F_\Gamma(x) - x$$

for $F_\Gamma(\cdot) \overset{\text{def}}{=} (I - \Gamma)x + \Gamma F(x)$. Note that the diagonal terms of Γ are bounded by 1. Then

$$||F_\Gamma(x) - F_\Gamma(y)||_\infty \leq \bar{\beta} ||x - y||_\infty \ \forall x, y,$$

where $\bar{\beta} \overset{\text{def}}{=} 1 - \epsilon(1 - \beta) \in (0,1)$. In particular, this is true for $\Gamma = \Lambda(t)$ for any $t \geq 0$. Thus once again a direct calculation shows that (7.4.5) holds and therefore $x(t) \to x^*$. (See the remark following Theorem 2 of

Chapter 10.) In fact, these observations extend to the situation when $\beta = 1$ as well. For the case when $\beta = 1$, existence of equilibrium is an *assumption* on $F(\cdot)$ and uniqueness need not hold. One can also consider 'weighted norms', such as $||x||_{\infty,w} \stackrel{\text{def}}{=} \sup_i w_i |x_i|$ for prescribed $w_i > 0, 1 \leq i \leq d$. We omit the details here as these will be self-evident after the developments in Chapter 10 where such 'fixed point solvers' are analyzed in greater detail.

Finally, note that replacing $a(\nu(i,n))$ in (7.1.2) by $a(n)$ would amount to distributed but *synchronous* iterations, as they presuppose a common clock. These can be analyzed along exactly the same lines, with the o.d.e. limit (7.1.5). In the case when $\{Y_n\}$ can be viewed as a controlled Markov chain on the power set Q of $\{1, 2, \ldots, d\}$, the analysis of sections 6.2 and 6.3 of Chapter 6 shows that the ith diagonal element of $\Lambda(t)$ will in fact be of the form $\sum_{i \in A \in Q} \pi_t(A)$, where π_t is the vector of stationary probabilities for this chain under *some* stationary policy. Note that in principle, $\{Y_n\}$ can always be cast as a controlled Markov chain on Q: Let the control space U be the set of probability measures on Q and let the controlled transition probability function be $p(j|i,u) = u(j)$ for $i, j \in Q, u \in U$. The control sequence is then the process of regular conditional laws of Y_{n+1} given $Y_k, k \leq n$, for $n \geq 0$. This gives a recipe for verifying (7.1.3) in many cases.

One may be able to 'rig' the stepsizes here so as to get the desired limiting o.d.e. For example, suppose the $\{Y_n\}$ above takes values in $\{\{i\} : 1 \leq i \leq d\}$, i.e., singletons alone. Suppose further that it is an ergodic Markov chain on this set. Suppose the chain has a stationary probability vector $[\pi_1, \ldots, \pi_d]$. Then by Corollary 8 of Chapter 6, $\Lambda(t) \equiv \text{diag}(\pi_1, \ldots, \pi_d)$. Thus if we use the stepsizes $\{a(n)/\pi_i\}$ for the ith component, we get the limiting o.d.e. $\dot{x}(t) = h(x(t))$ as desired. In practice, one may use $\{a(n)/\xi_n(i)\}$ instead, where $\xi_n(i)$ is an empirical estimate of π_i obtained by suitable averaging on a faster timescale so that it tracks π_i. (One could, for example, have $\xi_n(i) = \nu(i,n)/n$ if $na(n) \to 0$ as $n \to \infty$, i.e., the stepsizes $\{a(n)\}$ decrease slower than $\frac{1}{n}$, and therefore the analysis of section 6.1 applies.) This latter arrangement also extends in a natural manner to the more general case when $\{Y_n\}$ is 'controlled Markov'.

8

A Limit Theorem for Fluctuations

8.1 Introduction

To motivate the results of this chapter, consider the classical strong law of large numbers: Let $\{X_n\}$ be i.i.d. random variables with $E[X_n] = \mu, E[X_n^2] < \infty$. Let

$$S_0 = 0, \ S_n \stackrel{\text{def}}{=} \frac{\sum_{i=1}^n X_i}{n}, \ n \geq 1.$$

The strong law of large numbers (see, e.g., Section 4.2 of Borkar, 1995) states that

$$\frac{S_n}{n} \to \mu, \ \text{a.s.}$$

To cast this as a 'stochastic approximation' result, note that some simple algebraic manipulation leads to

$$
\begin{aligned}
S_{n+1} &= S_n + \frac{1}{n+1}(X_{n+1} - S_n) \\
&= S_n + \frac{1}{n+1}([\mu - S_n] + [X_{n+1} - \mu]) \\
&= S_n + a(n)(h(S_n) + M_{n+1})
\end{aligned}
$$

for

$$a(n) \stackrel{\text{def}}{=} \frac{1}{n+1}, \ h(x) \stackrel{\text{def}}{=} \mu - x \ \forall x, \ M_{n+1} \stackrel{\text{def}}{=} X_{n+1} - \mu.$$

In particular, $\{a(n)\}$ and $\{M_{n+1}\}$ are easily seen to satisfy the conditions stipulated for the stepsizes and martingale difference noise resp. in Chapter 2. Thus this is a valid stochastic approximation iteration. Its o.d.e. limit then is

$$\dot{x}(t) = \mu - x(t), \ t \geq 0,$$

88

which has μ as the unique globally asymptotically stable equilibrium. Its 'scaled limit' as in assumption (A5) of Chapter 3 is

$$\dot{x}(t) = -x(t), \ t \geq 0,$$

which has the origin as the unique globally asymptotically stable equilibrium. Thus by the theory developed in Chapter 2 and Chapter 3,

(i) the 'iterates' $\{S_n\}$ remain a.s. bounded, and

(ii) they a.s. converge to μ.

We have recovered the strong law of large numbers from stochastic approximation theory. Put differently, the a.s. convergence results for stochastic approximation iterations are nothing but a generalization of the strong law of large numbers for a class of dependent and not necessarily identically distributed random variables.

The classical strong law of large numbers, which states a.s. convergence of empirical averages to the mean, is accompanied by other limit theorems that quantify fluctuations around the mean, such as the central limit theorem, the law of iterated logarithms, the functional central limit theorem (Donsker's theorem), etc. It is then reasonable to expect similar developments for the stochastic approximation iterates. The aim of this chapter is to state a *functional central limit theorem* in this vein. This is proved in section 8.3, following some preliminaries in the next section. Section 8.4 specializes these results to the case when the iterates a.s. converge to a single deterministic limit and recovers the central limit theorem for stochastic approximation (see, e.g., Chung (1954), Fabian (1968)).

8.2 A tightness result

We shall follow the notation of section 4.3, which we briefly recall below. Thus our basic iteration in $\mathcal{R}^d$ is

$$x_{n+1} = x_n + a(n)(h(x_n) + M_{n+1}) \tag{8.2.1}$$

for $n \geq 0$, with the usual assumptions on $\{a(n)\}, \{M_{n+1}\}$, and the additional assumptions:

(A1) $h(\cdot)$ is continuously differentiable and both $h(\cdot)$ and the Jacobian matrix $\nabla h(\cdot)$ are uniformly Lipschitz.

(A2) $\frac{a(n)}{a(n+1)} \xrightarrow{n \uparrow \infty} 1$.

(A3) $\sup_n \|x_n\| < \infty$ a.s., $\sup_n E[\|x_n\|^4] < \infty$.

(A4) $\{M_n\}$ satisfy

$$E[M_{n+1}M_{n+1}^{\mathrm{T}}|M_i, x_i, i \leq n] = Q(x_n),$$
$$E[\|M_{n+1}\|^4|M_i, x_i, i \leq n] \leq K'(1 + \|x_n\|^4),$$

where $K' > 0$ is a suitable constant and $Q : \mathcal{R}^d \to \mathcal{R}^{d \times d}$ is a positive definite matrix-valued Lipschitz function such that the least eigenvalue of $Q(x)$ is bounded away from zero uniformly in x.

For the sake of simplicity, we also assume $a(n) \leq 1 \ \forall n$. As before, fix $T > 0$ and define $t(0) = 0$,

$$t(n) \overset{\text{def}}{=} \sum_{m=0}^{n-1} a(m),$$

$$m(n) \overset{\text{def}}{=} \min\{m \geq n : t(m) \geq t(n) + T\}, \ n \geq 1.$$

Thus $t(m(n)) \in [t(n) + T, t(n) + T + 1]$. For $n \geq 0$, let $x^n(t), t \geq t(n)$, denote the solution to

$$\dot{x}^n(t) = h(x^n(t)), \ t \geq t(n), \ x^n(t(n)) = x_n. \tag{8.2.2}$$

Then

$$x^n(t(j+1)) = x^n(t(j)) + a(j)\Big(h(x^n(t(j))) - \delta_j\Big), \tag{8.2.3}$$

where δ_j is the 'discretization error' as in Chapter 2, which is $O(a(j))$. Let

$$y_j \overset{\text{def}}{=} x_j - x^n(t(j)),$$

$$z_j \overset{\text{def}}{=} \frac{y_j}{\sqrt{a(j)}},$$

for $j \geq n, n \geq 0$. Subtracting (8.2.3) from (8.2.1) and using Taylor expansion, we have

$$y_{j+1} = y_j + a(j)(\nabla h(x^n(t(j)))y_j + \kappa_j + \delta_j) + a(j)M_{j+1}.$$

Here $\kappa_j = o(\|y_j\|)$ is the error in the Taylor expansion, which is also $o(1)$ in view of Theorem 2 of Chapter 2. Iterating, we have, for $0 \leq i \leq m(n) - n$,

$$
\begin{aligned}
y_{n+i} &= \Pi_{j=n}^{n+i-1}(1 + a(j)\nabla h(x^n(t(j))))y_n \\
&\quad + \sum_{j=n}^{n+i-1} a(j)\Pi_{k=j+1}^{n+i-1}(1 + a(k)\nabla h(x^n(t(k))))(\kappa_j + \delta_j + M_{j+1}) \\
&= \sum_{j=n}^{n+i-1} a(j)\Pi_{k=j+1}^{n+i-1}(1 + a(k)\nabla h(x^n(t(k))))(\kappa_j + \delta_j + M_{j+1}),
\end{aligned}
$$

because $y_n = 0$. Thus for i as above,

$$
\begin{aligned}
z_{n+i} = & \sum_{j=n}^{n+i-1} \sqrt{a(j)} \Pi_{k=j+1}^{n+i-1} (1 + a(k)\nabla h(x^n(t(k)))) \\
& \times \sqrt{\frac{a(k)}{a(k+1)}} M_{j+1} \sqrt{\frac{a(j)}{a(j+1)}} \\
& + \sum_{j=n}^{n+i-1} \sqrt{a(j)} \Pi_{k=j+1}^{n+i-1} (1 + a(k)\nabla h(x^n(t(k)))) \\
& \times \sqrt{\frac{a(k)}{a(k+1)}} (\kappa_j + \delta_j) \sqrt{\frac{a(j)}{a(j+1)}}.
\end{aligned}
\tag{8.2.4}
$$

Define $z^n(t), t \in I_n \stackrel{\text{def}}{=} [t(n), t(n) + T]$, by $z^n(t(j)) = z_j$, with linear interpolation on each $[t(j), t(j+1)]$ for $n \le j \le m(n)$. Let $\tilde{z}^n(t) = z^n(t(n)+t), t \in [0, T]$. We view $\{\tilde{z}^n(\cdot)\}$ as $C([0, T]; \mathcal{R}^d)$-valued random variables. Our first step will be to prove the tightness of their laws. For this purpose, we need the following technical lemma.

Let $\{X_n\}$ be a zero mean martingale w.r.t. the increasing σ-fields $\{\mathcal{F}_n\}$ with $X_0 = 0$ (say) and $\sup_{n \le N} E[|X_n|^4] < \infty$. Let $Y_n \stackrel{\text{def}}{=} X_n - X_{n-1}, n \ge 1$.

Lemma 1. *For a suitable constant $K > 0$,*

$$
E[\sup_{n \le N} |X_n|^4] \le K \left(\sum_{m=1}^{N} E[Y_m^4] + E[(\sum_{m=1}^{N} E[Y_m^2 | \mathcal{F}_{m-1}])^2] \right).
$$

Proof. For suitable constants $K_1, K_2, K_3, K_4 > 0$,

$$
\begin{aligned}
E[\sup_{n \leq N} |X_n|^4] &\leq K_1 E[(\sum_{m=1}^{N} Y_m^2)^2] \\
&\leq K_2 (E[(\sum_{m=1}^{N} (Y_m^2 - E[Y_m^2 | \mathcal{F}_{m-1}]))^2] \\
&\quad + E[(\sum_{m=1}^{N} E[Y_m^2 | \mathcal{F}_{m-1}])^2]) \\
&\leq K_3 (E[\sum_{m=1}^{N} (Y_m^2 - E[Y_m^2 | \mathcal{F}_{m-1}])^2] \\
&\quad + E[(\sum_{m=1}^{N} E[Y_m^2 | \mathcal{F}_{m-1}])^2]) \\
&\leq K_4 (E[\sum_{m=1}^{N} Y_m^4] \\
&\quad + E[(\sum_{m=1}^{N} E[Y_m^2 | \mathcal{F}_{m-1}])^2]),
\end{aligned}
$$

where the first and the third inequalities follow from Burkholder's inequality (see Appendix C). ∎

Applying Lemma 1 to (8.2.4), we have:

Lemma 2. *For $m(n) \geq \ell > k \geq n, n \geq 0$,*

$$
\begin{aligned}
E[\| z^n(t(\ell)) - z^n(t(k)) \|^4] &= O((\sum_{j=k}^{\ell} a(j))^2) \\
&= O(|t(\ell) - t(k)|^2).
\end{aligned}
$$

Proof. Let ℓ, k be as above. By (A2), $\sqrt{a(j)/a(j+1)}$ is uniformly bounded in j. Since h is uniformly Lipschitz, ∇h is uniformly bounded and thus for $n \leq k < m(n)$,

$$
\begin{aligned}
\| \Pi_{r=k+1}^{\ell} (1 + a(r)\nabla h(x^n(t(r)))) \| &\leq e^{K_1 \sum_{r=k+1}^{\ell} a(r)} \\
&\leq e^{(T+1)K_1},
\end{aligned}
$$

for a suitable bound $K_1 > 0$ on $\| \nabla h(\cdot) \|$. Also, for any $\eta > 0$, $\| \kappa_j \| \leq \eta \| y_j \| = \sqrt{a(j)} \eta \| z_j \|$ for sufficiently large j. Hence we have for large j and

$\eta' = \eta(\sup_n \sqrt{\frac{a(n)}{a(n+1)}})^2,$

$\| \sum_{j=k+1}^{\ell} \sqrt{a(j)} \Pi_{r=j}^{\ell} (1 + a(r)\nabla h(x^n(t(r)))) \sqrt{\frac{a(r)}{a(r+1)}} \kappa_j \sqrt{\frac{a(j)}{a(j+1)}} \|$

$$\leq \eta' e^{K_1(T+1)} \sum_{j=k+1}^{\ell} a(j)\|z_j\|.$$

Similarly, since $\|\delta_j\| \leq K'a(j)$ for some $K' > 0$,

$\| \sum_{j=k+1}^{\ell} \sqrt{a(j)} \Pi_{r=j}^{\ell} (1 + a(r)\nabla h(x^n(t(r)))) \sqrt{\frac{a(r)}{a(r+1)}} \delta_j \sqrt{\frac{a(j)}{a(j+1)}} \|$

$$\leq K e^{K_1(T+1)} \sum_{j=k+1}^{\ell} a(j)^{\frac{3}{2}}$$

for some constant $K > 0$. Applying the above lemma to $\Psi_k \stackrel{\text{def}}{=}$ the first term on the right-hand side of (8.2.4) with $n + i = k$, $E[\sup_{n \leq m \leq k} \|\Psi_m\|^4]$ is seen to be bounded by

$$K_1((\sum_{j=n}^{k} a(j))^2 + \sum_{j=n}^{k} a(j)^2)$$

for some $K_1 > 0$, where we have used the latter parts of (A3) and (A4). Combining this with the foregoing, we have

$$E[\|z^n(t(k))\|^4] \leq K_2\Big((\sum_{j=n}^{k} a(j))^2 + \sum_{j=n}^{k} a(j)^2$$
$$+ (\sum_{j=n}^{k} a(j)^{\frac{3}{2}})^4 + E[(\sum_{j=n}^{k} a(j)\|z_j\|)^4]\Big)$$

for a suitable $K_2 > 0$. Since $z_j = z^n(t(j))$, $a(j) \leq 1$ and $\sum_{j=n}^{m(n)} a(j) \leq T + 1$, we have

$$E[(\sum_{j=n}^{k} a(j)\|z_j\|)^4] \leq (\sum_{j=n}^{k} a(j))^4 E[\Big(\frac{1}{\sum_{j=n}^{k} a(j)} \sum_{j=n}^{k} a(j)\|z_j\|\Big)^4]$$

$$\leq (T+1)^3 \sum_{j=n}^{k} a(j)E[\|z_j\|^4] \tag{8.2.5}$$

by the Jensen inequality. Also, $\sum_{j=n}^{k} a(j)^2 \leq (\sum_{j=n}^{k} a(j))^2 \leq (T+1)^2$, $\sum_{j=n}^{k} a(j)^{\frac{3}{2}} \leq (T+1)\sup_j \sqrt{a(j)}$. Thus, for a suitable $K_3 > 0$,

$$E[\|z^n(t(k))\|^4] \leq K_3(1 + \sum_{j=n}^{k} a(j)E[\|z^n(t(j))\|^4]).$$

By the discrete Gronwall inequality, it follows that

$$\sup_{n \leq k \leq m(n)} E[\|z^n(t(k))\|^4] \leq K_4 < \infty \tag{8.2.6}$$

for a suitable $K_4 > 0$. Arguments analogous to (8.2.5) then also lead to

$$E[(\sum_{j=k}^{\ell} a(j)\|z_j\|)^4] \le (\sum_{j=k}^{\ell} a(j))^3 K_4.$$

Thus

$$
\begin{aligned}
E[\|z^n(t(\ell)) - z^n(t(k))\|^4] &\le K_2\Big((\sum_{j=k}^{\ell} a(j))^2 + \sum_{j=k}^{\ell} a(j)^2 \\
&\quad + (\sum_{j=k}^{\ell} a(j)^{\frac{3}{2}})^4 + (\sum_{j=k}^{\ell} a(j))^3 K_4\Big) \\
&\le K_5(\sum_{j=k}^{\ell} a(j))^2 \\
&= O(|t(\ell) - t(k)|^2).
\end{aligned}
$$

∎

A small variation of the argument used to prove (8.2.6) shows that

$$E[\sup_{n \le k \le \ell} \|z^n(t(k))\|^4] \le K'(1 + \sum_{m=n}^{\ell} E[\sup_{n \le k \le m} \|z^n(t(k))\|^4]),$$

which, by the discrete Gronwall inequality, improves (8.2.6) to

$$E[\sup_{n \le k \le m(n)} \|z^n(t(k))\|^4] \le K_5 < \infty. \tag{8.2.7}$$

We shall use this bound later. A claim analogous to Lemma 2 holds for $x^n(\cdot)$:

Lemma 3. *For $t(n) \le s < t \le t(n) + T$,*

$$E[\|x^n(t) - x^n(s)\|^4] \le K(T)|t - s|^2$$

for a suitable constant $K(T) > 0$ depending on T.

Proof. Since $h(\cdot)$ is Lipschitz,

$$\|h(x)\| \le K(1 + \|x\|) \tag{8.2.8}$$

for some constant $K > 0$. Since

$$\sup_n E[\|x^n(t(n))\|^4] = \sup_n E[\|x_n\|^4] < \infty$$

by (A3), a straightforward application of the Gronwall inequality in view of (8.2.8) leads to

$$\sup_{t \in [0,T]} E[\|x^n(t(n) + t)\|^4] < \infty.$$

Thus for some $K', K(T) > 0$,

$$
\begin{aligned}
E[||x^n(t) - x^n(s)||^4] &\leq E[|| \int_s^t h(x^n(y))dy||^4] \\
&\leq (t-s)^4 E[||\frac{1}{t-s} \int_s^t h(x^n(y))dy||^4] \\
&\leq (t-s)^3 E[\int_s^t ||h(x^n(y))||^4 dy] \\
&\leq (t-s)^3 E[\int_s^t K'(1 + ||x^n(y)||)^4 dy] \\
&\leq K(T)|t-s|^2,
\end{aligned}
$$

which is the desired bound. ∎

We shall need the following well-known criterion for tightness of probability measures on $C([0,T]; \mathcal{R}^d)$:

Lemma 4. *Let $\{\xi_\alpha(\cdot)\}$, for α belonging to some prescribed index set J, be a family of $C([0,T]; \mathcal{R}^d)$-valued random variables such that the laws of $\{\xi_\alpha(0)\}$ are tight in $\mathcal{P}(\mathcal{R}^d)$, and for some constants $a, b, c > 0$*

$$
E[||\xi_\alpha(t) - \xi_\alpha(s)||^a] \leq b|t-s|^{1+c} \ \forall \ \alpha \in J, \ t, s \in [0,T]. \tag{8.2.9}
$$

Then the laws of $\{\xi_\alpha(\cdot)\}$ are tight in $\mathcal{P}(C([0,T]; \mathcal{R}^d))$.

See Billingsley (1968, p. 95) for a proof.

Let $\tilde{x}^n(t) = x^n(t(n) + t), t \in [0,T]$. Then we have:

Lemma 5. *The laws of the processes $\{(\tilde{z}^n(\cdot), \tilde{x}^n(\cdot)), n \geq 0\}$ are relatively compact in $\mathcal{P}(C([0,T]; \mathcal{R}^d))^2$.*

Proof. Note that $\tilde{z}^n(0) = 0 \ \forall n$ and hence have trivially tight laws. Tightness of the laws of $\{\tilde{z}^n(\cdot), n \geq 0\}$ then follows by combining Lemmas 2 and 4 above. Tightness of the laws of $\{\tilde{x}(0)\}$ follows from the second half of (A3), as

$$
\begin{aligned}
P(||x^n(t)|| > a) &\leq \frac{E||x^n(t)||^4}{a^4} \\
&\leq \frac{\bar{K}}{a^4},
\end{aligned}
$$

for a suitable constant $\bar{K} > 0$. Tightness of the laws of $\{\tilde{x}^n(\cdot)\}$ then follows by Lemmas 3 and 4. Since tightness of marginals implies tightness of joint laws, tightness of the joint laws of $\{(\tilde{z}^n(\cdot), \tilde{x}^n(\cdot))\}$ follows. The claim is now immediate from Prohorov's theorem (see Appendix C). ∎

In view of this lemma, we may take a subsequence of $\{(\tilde{z}^n(\cdot), \tilde{x}^n(\cdot))\}$ that converges in law to a limit (say) $\{(z^*(\cdot), x^*(\cdot))\}$. Denote this subsequence again by $\{(\tilde{z}^n(\cdot), \tilde{x}^n(\cdot))\}$ by abuse of notation. In the next section we characterize this limit.

8.3 The functional central limit theorem

To begin with, we shall invoke Skorohod's theorem (see Appendix C) to suppose that

$$(\tilde{z}^n(\cdot), \tilde{x}^n(\cdot)) \to (z^*(\cdot), x^*(\cdot)), \text{ a.s.} \tag{8.3.1}$$

in $C([0, T]; \mathcal{R}^d)^2$. Since the trajectories of the o.d.e. $\dot{x}(t) = h(x(t))$ form a closed set in $C([0, T]; \mathcal{R}^d)$ (see Appendix A), it follows that $x^*(\cdot)$ satisfies this o.d.e. In fact, we know separately from the developments of Chapter 2 that it would be in an internally chain transitive invariant set thereof. To characterize $z^*(\cdot)$, it is convenient to work with

$$
\begin{aligned}
z_{j+1} = {} & \sqrt{\frac{a(j)}{a(j+1)}} z_j + a(j) \nabla h(x^n(j)) \sqrt{\frac{a(j)}{a(j+1)}} z_j \\
& + \sqrt{a(j)} \sqrt{\frac{a(j)}{a(j+1)}} M_{j+1} + o(a(j)),
\end{aligned}
$$

for $n \le j \le m(n)$, which lead to

$$
\begin{aligned}
z_{j+1} = {} & z_n + \sum_{k=n}^{j} \left(\sqrt{\frac{a(k)}{a(k+1)}} - 1 \right) z_k \\
& + \sum_{k=n}^{j} a(k) \sqrt{\frac{a(k)}{a(k+1)}} \nabla h(x^n(t(k)) z_k \\
& + \sum_{k=n}^{j} \sqrt{a(k)} \sqrt{\frac{a(k)}{a(k+1)}} M_{k+1} + o(1),
\end{aligned}
$$

for j in the above range. Thus

$$
\begin{aligned}
z^n(t(j+1)) = {} & z^n(t(n)) + (\zeta_j - \zeta_n) \\
& + \int_{t(n)}^{t(j+1)} \nabla h(x^n(y)) z^n(y) b(y) dy \\
& + \sum_{k=n}^{j} \sqrt{a(k)} \sqrt{\frac{a(k)}{a(k+1)}} M_{k+1} + o(1),
\end{aligned}
$$

where:

- $\zeta_m = \sum_{i=n}^{m} \left(\sqrt{\frac{a(i)}{a(i+1)}} - 1 \right) z_i$, and

- $b(t(j)) \overset{\text{def}}{=} \sqrt{a(j)/a(j+1)}, j \geq 0$, with linear interpolation on each interval $[t(j), t(j+1)]$.

Note that

$$b(t) \overset{t\uparrow\infty}{\to} 1, \tag{8.3.2}$$

and, for $t(\ell) = \min\{t(k) : t(k) \geq t(n) + T\}$,

$$\max_{n \leq j \leq \ell} \|\zeta_j - \zeta_n\| \leq \sup_{i \geq n} |\sqrt{a(i)/a(i+1)} - 1| \sup_{t \in [t(n), t(\ell)]} \|z^n(t)\|(T+1). \tag{8.3.3}$$

By (8.2.7) and (A2), the right-hand side tends to zero in fourth moment and therefore in law as $n \to \infty$ for any $T > 0$. Fix $t > s$ in $[0, T]$ and let $g \in C_b(C([0, s]; \mathcal{R}^d)^2)$. Then, since $\{M_k\}$ is a martingale difference sequence, we have

$$\|E[(\tilde{z}^n(t) - \tilde{z}^n(s) - \int_s^t \nabla h(\tilde{x}^n(y))b(t(n) + y)\tilde{z}^n(y)dy)$$
$$\times g(\tilde{z}^n([0, s]), \tilde{x}^n([0, s]))]\|$$
$$= o(1).$$

Here we use the notation $f([0, s])$ to denote the trajectory segment $f(y), 0 \leq y \leq s$. Letting $n \to \infty$, we then have, in view of (8.3.2),

$$E[(z^*(t) - z^*(s) - \int_s^t \nabla h(x^*(y))z^*(y)dy)g(z^*([0, s]), x^*([0, s]))] = 0.$$

Letting $\mathcal{G}_t \overset{\text{def}}{=}$ the completion of $\cap_{s \geq t}\sigma(z^*(y), x^*(y), y \leq s)$ for $t \geq 0$, it then follows by a standard monotone class argument that

$$z^*(t) - \int_0^t \nabla h(x^*(s))z^*(s)ds, \ t \in [0, T],$$

is a martingale.

For $t \in [0, T]$, define $\Sigma^n(t)$ by

$$\Sigma^n(t(j) - t(n)) = \sum_{k=n}^j a(k)b(t(k))^2 Q(x^n(t(k))),$$

for $n \leq j \leq m(n)$, with linear interpolation on each $[t(j) - t(n), t(j+1) - t(n)]$. Then

$$\sum_{k=n}^j a(k)b(k)^2 M_{j+1} M_{j+1}^{\mathrm{T}} - \Sigma^n(t(j) - t(n)), \ n \leq j \leq m(n),$$

is a martingale by (A4). Therefore for t, s as above and

$$q^n(t) \overset{\text{def}}{=} \tilde{z}^n(t) - \int_0^t \nabla h(\tilde{x}^n(y))b(t(n) + y)\tilde{z}^n(y)dy, \ t \in [0, T],$$

we have

$$\|E[(q^n(t)q^n(t)^T - q^n(s)q^n(s)^T - (\Sigma^n(t) - \Sigma^n(s)))$$
$$\times g(\tilde{z}^n([0,s]), \tilde{x}^n([0,s]))]\|$$
$$= o(1).$$

(The multiplication by $g(\cdots)$ and the expectation are componentwise.) Passing to the limit as $n \to \infty$, one concludes as before that

$$\left(z^*(t) - \int_0^t \nabla h(x^*(s))z^*(s)ds\right)\left(z^*(t) - \int_0^t \nabla h(x^*(s))z^*(s)ds\right)^T$$
$$- \int_0^t Q(x^*(s))ds$$

is a $\{\mathcal{G}_t\}$-martingale for $t \in [0,T]$. From the results of Wong (1971), it then follows that on a possibly augmented probability space, there exists a d-dimensional Brownian motion $B(t), t \geq 0$, such that

$$z^*(t) = \int_0^t \nabla h(x^*(s))z^*(s)ds + \int_0^t D(x^*(s))dB(s), \qquad (8.3.4)$$

where $D(x) \in \mathcal{R}^{d \times d}$ for $x \in \mathcal{R}^d$ is a positive semidefinite, Lipschitz (in x), square-root of the matrix $Q(x)$.

Remarks: (1) A square-root as above always exists under our hypotheses, as shown in Theorem 5.2.2 of Stroock and Varadhan (1979).
(2) Equation (8.3.4) specifies $z^*(\cdot)$ as a solution of a *linear* stochastic differential equation. A 'variation of constants' argument leads to the explicit expression

$$z^*(t) = \int_0^t \Phi(t,s)Q(x^*(s))dB(s), \qquad (8.3.5)$$

where $\Phi(t,s), t \geq s \geq 0$, satisfies the linear matrix differential equation

$$\frac{d}{dt}\Phi(t,s) = \nabla h(x^*(t))\Phi(t,s), \quad t \geq s; \quad \Phi(s,s) = I_d. \qquad (8.3.6)$$

Here I_d denotes the $d \times d$ identity matrix. In particular, if $x^*(\cdot)$ were a deterministic trajectory, then by (8.3.6), $\Phi(\cdot,\cdot)$ would be deterministic too and by (8.3.5), $z^*(\cdot)$ would be the solution of a linear stochastic differential equation with deterministic coefficients and zero initial condition. In particular, (8.3.5) would then imply that it is a zero mean Gaussian process.

Summarizing, we have:

Theorem 6. *The limits in law $(z^*(\cdot), x^*(\cdot))$ of $\{(\tilde{z}^n(\cdot), \tilde{x}^n(\cdot))\}$ are such that $x^*(\cdot)$ is a solution of the o.d.e. (9.1.2) belonging to an internally chain transitive invariant set thereof, and $z^*(\cdot)$ satisfies (8.3.4).*

8.4 The convergent case

We now consider the special case when the o.d.e. $\dot{x}(t) = h(x(t))$ has a unique globally asymptotically stable equilibrium $\hat{x}$. Then under our conditions, $x_n \to \hat{x}$ a.s. as $n \uparrow \infty$ by Theorem 2 of Chapter 2. We may also suppose that all eigenvalues of $\nabla h(\hat{x})$ have strictly negative real parts. Then $x^*(\cdot) \equiv \hat{x}$ and thus (8.3.4) reduces to the *constant coefficient* linear stochastic differential equation

$$z^*(t) = \int_0^t \nabla h(\hat{x}) z^*(s) ds + \int_0^t D(\hat{x}) dB(s),$$

leading to

$$z^*(t) = \int_0^t e^{\nabla h(\hat{x})(t-s)} D(\hat{x}) dB(s).$$

In particular, $z^*(t)$ is a zero mean Gaussian random variable with covariance matrix given by

$$\Gamma(t) \overset{\text{def}}{=} \int_0^t e^{\nabla h(\hat{x})(t-s)} Q(\hat{x}) e^{\nabla h(\hat{x})^{\mathrm{T}}(t-s)} ds$$

$$= \int_0^t e^{\nabla h(\hat{x})u} Q(\hat{x}) e^{\nabla h(\hat{x})^{\mathrm{T}}u} du,$$

after a change of variable $u = t - s$. Thus, as $t \to \infty$, the law of $z^*(t)$ converges to the stationary distribution of this Gauss–Markov process, which is zero mean Gaussian with covariance matrix

$$\Gamma^* \overset{\text{def}}{=} \lim_{t \to \infty} \Gamma(t)$$

$$= \int_0^\infty e^{\nabla h(\hat{x})s} Q(\hat{x}) e^{\nabla h(\hat{x})^{\mathrm{T}}s} ds.$$

Note that

$$\nabla h(\hat{x})\Gamma^* + \Gamma^* \nabla h(\hat{x})^{\mathrm{T}} = \lim_{t \to \infty} (\nabla h(\hat{x})\Gamma(t) + \Gamma(t)\nabla h(\hat{x})^{\mathrm{T}})$$

$$= \lim_{t \to \infty} \int_0^t \frac{d}{du} \left(e^{\nabla h(\hat{x})u} Q(\hat{x}) e^{\nabla h(\hat{x})^{\mathrm{T}}u} \right) du$$

$$= \lim_{t \to \infty} (e^{\nabla h(\hat{x})t} Q(\hat{x}) e^{\nabla h(\hat{x})^{\mathrm{T}}t} - Q(\hat{x}))$$

$$= -Q(\hat{x}),$$

in view of our assumption on $\nabla h(\hat{x})$ above. Thus Γ^* satisfies the matrix equation

$$\nabla h(\hat{x})\Gamma^* + \Gamma^* \nabla h(\hat{x})^{\mathrm{T}} + Q(\hat{x}) = 0. \qquad (8.4.1)$$

From the theory of linear systems of differential equations, it is well-known that Γ^* is the unique positive definite solution to the 'Liapunov equation' (8.4.1) (see, e.g., Kailath, 1980, p. 179).

Since the Gaussian density with zero mean and covariance matrix $\Gamma(t)$ converges pointwise to the Gaussian density with zero mean and covariance matrix Γ^*, it follows from Scheffé's theorem (see Appendix C) that the law of $z^*(t)$ tends to the stationary distribution in total variation and hence in $\mathcal{P}(\mathcal{R}^d)$. Let $\epsilon > 0$ and pick T above large enough that for $t = T$, the two are at most ϵ apart with respect to a suitable metric ρ compatible with the topology of $\mathcal{P}(\mathcal{R}^d)$. It then follows that the law of $\tilde{z}^n(T)$ converges to the ϵ-neighbourhood (w.r.t. ρ) of the stationary distribution as $n \uparrow \infty$. Since $\epsilon > 0$ was arbitrary, it follows that it converges in fact to this distribution. We have thus proved the '*Central Limit Theorem*' for stochastic approximation:

Theorem 7. *The law of z_n converges to the Gaussian distribution with zero mean and covariance matrix Γ^* given by the unique positive definite solution to (8.4.1).*

One important implication of this result is the following: it suggests that in a certain sense, the convergence rate of x_n to $\hat{x}$ is $O(\sqrt{a(n)})$. Also, one can read off 'confidence intervals' for finite runs based on Gaussian approximation, see Hsieh and Glynn (2000).

We have presented a simple case of the functional central limit theorem for stochastic approximation. A much more general statement that allows for a 'Markov noise' on the natural timescale as in Chapter 6 is available in Benveniste, Metivier and Priouret (1990). In addition, there are also other limit theorems available for stochastic approximation iterations, such as convergence rates for moments (Gerencsér, 1992), a 'pathwise central limit theorem' for certain scaled empirical measures (Pelletier, 1999), a law of iterated logarithms (Pelletier, 1998), Strassen-type strong invariance principles (Lai and Robbins, 1978; Pezeshki-Esfahani and Heunis, 1997), and Freidlin – Wentzell type 'large deviations' bounds (Dupuis, 1988; Dupuis and Kushner, 1989).

9

Constant Stepsize Algorithms

9.1 Introduction

In many practical circumstances, it is more convenient to use a small constant stepsize $a(n) \equiv a \in (0,1)$ rather than the decreasing stepsize considered thus far. One such situation is when the algorithm is 'hard-wired' and decreasing stepsize may mean additional overheads. Another important scenario is when the algorithm is expected to operate in a slowly varying environment (e.g., in tracking applications) where it is important that the timescale of the algorithm remain reasonably faster than the timescale on which the environment is changing, for otherwise it would never adapt.

Naturally, for constant stepsize one has to forgo the strong convergence statements we have been able to make for decreasing stepsizes until now. A rule of thumb, to be used with great caution, is that in the passage from decreasing to small positive stepsize, one replaces a '*converges a.s. to*' statement with a '*concentrates with a high probability in a neighbourhood of*' statement. We shall make this more precise in what follows, but the reason for thus relaxing the claims is not hard to guess. Consider for example the iteration

$$x_{n+1} = x_n + a[h(x_n) + M_{n+1}], \tag{9.1.1}$$

where $\{M_n\}$ are i.i.d. with Gaussian densities. Suppose the o.d.e.

$$\dot{x}(t) = h(x(t)) \tag{9.1.2}$$

has a unique globally asymptotically stable equilibrium x^*. Observe that $\{x_n\}$ is then a Markov process and if it is stable (i.e., the laws of $\{x_n\}$ remain *tight* – see Appendix C) then the best one can hope for is that it will have a stationary distribution which assigns a high probability to a neighbourhood of x^*. On the other hand, because of additive Gaussian noise, the stationary distribution will have full support. Using the well-known recurrence properties

of such Markov processes, it is not hard to see that both 'sup$_n ||x_n|| < \infty$ *a.s.*'
and '$x_n \to x^*$ *a.s.*' are untenable, because $\{x_n\}$ will visit any given open set
infinitely often with probability one.

In this chapter, we present many counterparts of the results thus far for
constant stepsize. The treatment will be rather sketchy, emphasizing mainly
the points of departures from the diminishing stepsizes. What follows also
extends more generally to bounded stepsizes.

9.2 Asymptotic behaviour

We shall assume (A1), (A3) of Chapter 2 and replace (A4) there by

$$C \overset{\text{def}}{=} \sup_n E[||x_n||^2]^{\frac{1}{2}} < \infty \tag{9.2.1}$$

and

$$\sup_n E[G(||x_n||^2)] < \infty \tag{9.2.2}$$

for some $G : [0, \infty) \to [0, \infty)$ satisfying $G(t)/t \overset{t\uparrow\infty}{\to} \infty$. Condition (9.2.2)
is equivalent to the statement that $\{||x_n||^2\}$ are uniformly integrable – see,
e.g., Theorem 1.3.4, p. 10, of Borkar (1995). Let $L > 0$ denote the Lipschitz
constant of h as before. As observed earlier, its Lipschitz continuity implies at
most linear growth. Thus we have the equivalent statements

$$||h(x)|| \leq K_1(1 + ||x||) \ \ \text{or} \ \ K_2\sqrt{1 + ||x||^2},$$

for suitable $K_1, K_2 > 0$. We may use either according to convenience. Imitating
the developments of Chapter 2 for decreasing stepsizes, let $t(n) = na, n \geq 0$.
Define $\bar{x}(\cdot)$ by $\bar{x}(t(n)) = x_n \ \forall n$ with $\bar{x}(t)$ defined on $[t(n), t(n+1)]$ by linear
interpolation for all n, so that it is a piecewise linear, continuous function. As
before, let $x^s(t), t \geq s$, denote the trajectory of (9.1.2) with $x^s(s) = \bar{x}(s)$. In
particular,

$$x^s(t) = \bar{x}(s) + \int_s^t h(x^s(y))dy$$

for $t \geq s$ implies that

$$||x^s(t)|| \leq ||\bar{x}(s)|| + \int_s^t K_1(1 + ||x^s(y)||)dy.$$

By the Gronwall inequality, we then have

$$||x^s(t)|| \leq K_\tau(1 + ||\bar{x}(s)||), \ t \in [s, s+\tau], \ \tau > 0.$$

for a $K_\tau > 0$. In turn this implies

$$||h(x^s(t))|| \leq K_1(1 + K_\tau(1 + ||\bar{x}(s)||)) \leq \Delta(\tau)(1 + ||\bar{x}(s)||), \ t \in [s, s+\tau],$$

for $\tau > 0, \Delta(\tau) \stackrel{\text{def}}{=} K_1(1 + K_\tau)$. Thus for $t > t'$ in $[s, s + \tau]$,

$$
\begin{aligned}
\|x^s(t) - x^s(t')\| &\leq \int_{t'}^t \|h(x^s(y)) - h(x^s(s))\| dy \\
&\leq \Delta(\tau)(1 + \|\bar{x}(s)\|)(t - t').
\end{aligned} \tag{9.2.3}
$$

The key estimate for our analysis is:

Lemma 1. *For any $T > 0$,*

$$
E[\sup_{t \in [0,T]} \|\bar{x}(s + t) - x^s(s + t)\|^2] = O(a). \tag{9.2.4}
$$

Proof. Consider $T = Na$ for some $N > 0$. For $t \geq 0$, let $[t] \stackrel{\text{def}}{=} \max\{na : n \geq 0, na \leq t\}$. Let $\zeta_n \stackrel{\text{def}}{=} a \sum_{m=1}^n M_m, n \geq 1$. Then for $n \geq 0$ and $1 \leq m \leq N$, we have

$$
\bar{x}(t(n + m)) = \bar{x}(t(n)) + \int_{t(n)}^{t(n+m)} h(\bar{x}([t]))dt + (\zeta_{m+n} - \zeta_n).
$$

Also,

$$
\begin{aligned}
x^{t(n)}(t(n + m)) &= \bar{x}(t(n)) + \int_{t(n)}^{t(n+m)} h(x^{t(n)}([t]))dt \\
&\quad + \int_{t(n)}^{t(n+m)} (h(x^{t(n)}(t)) - h(x^{t(n)}([t])))dt.
\end{aligned} \tag{9.2.5}
$$

Recall that L is a Lipschitz constant for $h(\cdot)$. Clearly,

$$
\begin{aligned}
\| \int_{t(n)}^{t(n+m)} &(h(\bar{x}([t])) - h(x^{t(n)}([t])))dt\| \\
&= a\| \sum_{k=0}^{m-1} \left(h(\bar{x}(t(n + k))) - h(x^{t(n)}(t(n + k))) \right) \| \\
&\leq aL \sum_{k=0}^{m-1} \|\bar{x}(t(n + k)) - x^{t(n)}(t(n + k))\| \\
&\leq aL \sum_{k=0}^{m-1} \sup_{j \leq k} \|\bar{x}(t(n + j)) - x^{t(n)}(t(n + j))\|.
\end{aligned} \tag{9.2.6}
$$

By (9.2.3), we have

$$\left\| \int_{t(n+m)}^{t(n+m+1)} (h(x(t)) - h(x([t]))) dt \right\|$$

$$\leq \int_{t(n+m)}^{t(n+m+1)} \|h(x(t)) - h(x([t]))\| dt$$

$$\leq L \int_{t(n+m)}^{t(n+m+1)} \|x(t) - x([t])\| dt$$

$$\leq \frac{1}{2} a^2 L \Delta(Na)(1 + \|\bar{x}(t(n))\|)$$

$$\stackrel{\text{def}}{=} \frac{1}{2} a^2 \hat{K}(1 + \|\bar{x}(t(n))\|). \tag{9.2.7}$$

Subtracting (9.2.5) from (9.2), we have, by (9.2.6) and (9.2.7),

$$\sup_{0 \leq k \leq m} \|\bar{x}(t(n+k)) - x^{t(n)}(t(n+k)))\|$$

$$\leq aL \sum_{k=0}^{m-1} \sup_{0 \leq j \leq k} \|\bar{x}(t(n+j)) - x^{t(n)}(t(n+j))\|$$

$$+ aT\hat{K}(1 + \|\bar{x}(t(n))\|) + \sup_{1 \leq j \leq m} \|\zeta_{n+j} - \zeta_n\|.$$

By Burkholder's inequality (see Appendix C) and assumption (A3) of Chapter 2,

$$E[\sup_{1 \leq j \leq m} \|\zeta_{n+j} - \zeta_n\|^2] \leq a^2 K E[\sum_{0 \leq j < m} \|M_{n+j}\|^2]$$

$$\leq a^2 \tilde{K} \sum_{0 \leq j < m} (1 + E[\|\bar{x}(t(n+j))\|^2])$$

$$\leq a^2 \tilde{K} N (1 + C^2)$$

$$= a\tilde{K} T (1 + C^2)$$

for $m \leq N$, C as in (9.2.1) and suitable $K, \tilde{K} > 0$. Hence

$$E[\sup_{k \leq m} \|\bar{x}(t(n+k)) - x^{t(n)}(t(n+k)))\|^2]^{\frac{1}{2}}$$

$$\leq aL \sum_{k=0}^{m-1} E[\sup_{j \leq k} \|\bar{x}(t(n+j)) - x^{t(n)}(t(n+j))\|^2]^{\frac{1}{2}}$$

$$+ aT\hat{K} E[(1 + \|\bar{x}(t(n))\|)^2]^{\frac{1}{2}} + E[\sup_{1 \leq j \leq m} \|\zeta_{n+m} - \zeta_n\|^2]^{\frac{1}{2}}$$

$$\leq aL \sum_{k=0}^{m-1} E[\sup_{j \leq k} \|\bar{x}(t(n+j)) - x^{t(n)}(t(n+j))\|^2]^{\frac{1}{2}}$$

$$+ aTK\sqrt{1 + C^2} + \sqrt{a\tilde{K} T (1 + C^2)}$$

for a suitable $K > 0$. By the discrete Gronwall inequality (Appendix B), it follows that

$$E[\sup_{n \leq j \leq n+N} \|\bar{x}(t(j)) - x^{t(n)}(t(j))\|^2]^{\frac{1}{2}} \leq \sqrt{a}\bar{K}$$

for a suitable $\bar{K} > 0$ that depends on T. Since

$$E[\sup_{t \in [t(k), t(k+1)]} \|\bar{x}(t) - \bar{x}(t(k))\|^2]^{\frac{1}{2}}$$

and

$$E[\sup_{t \in [t(k), t(k+1)]} \|x^{t(n)}(t) - x^{t(n)}(t(k))\|^2]^{\frac{1}{2}}$$

are also $O(\sqrt{a})$, it is easy to deduce from the above that

$$E[\sup_{t \in [0,T]} \|\bar{x}(t(n) + t) - x^{t(n)}(t(n) + t)\|^2]^{\frac{1}{2}} \leq \sqrt{a}K$$

for a suitable $K > 0$. This completes the proof for T of the form $T = Na$. The general case can be proved easily from this. ∎

Let (9.1.2) have a globally asymptotically stable compact attractor A and let $\rho(x, A) \overset{\text{def}}{=} \min_{y \in A} \|x - y\|$ denote the distance of $x \in \mathcal{R}^d$ from A. For purposes of the proof of the next result, we introduce $R > 0$ such that

(i) $A^a \overset{\text{def}}{=} \{x \in \mathcal{R}^d : \rho(x, A) \leq a\} \subset B(R) \overset{\text{def}}{=} \{x \in \mathcal{R}^d : \|x\| < R\}$, and
(ii) for $a > 0$ as above,

$$\sup_n P(\|x_n\| \geq R) < a \text{ and } \sup_n E[\|x_n\|^2 I\{\|x_n\| \geq R\}] < a. \quad (9.2.8)$$

(The second part of (9.2.8) is possible by the uniform integrability condition (9.2.2).)

By global asymptotic stability of (9.1.2), we may pick $T = Na > 0$ large enough such that for any solution $x(\cdot)$ thereof with $x(0) \in B(R) - A$, one has $x(T) \in A^a$ and

$$\rho(x(T), A) \leq \frac{1}{2}\rho(x(0), A). \quad (9.2.9)$$

We also need the following lemma:

Lemma 2. *There exists a constant $K^* > 0$ depending on T above, such that for $t \geq 0$,*

$$E[\rho(\bar{x}(t + T), A)^2 I\{\bar{x}(t) \in A^a\}]^{\frac{1}{2}} \leq K^* \sqrt{a},$$
$$E[\rho(\bar{x}(t + T), A)^2 I\{\bar{x}(t) \in B(R)^c\}]^{\frac{1}{2}} \leq K^* \sqrt{a}.$$

Proof. In what follows, $K^* > 0$ denotes a suitable constant possibly depending on T, not necessarily the same each time. For $x^t(\cdot)$ as above, we have

$$E[\rho(\bar{x}(t+T), A)^2 I\{\bar{x}(t) \in A^a\}]^{\frac{1}{2}}$$
$$\leq \quad E[\rho(x^t(t+T), A)^2 I\{\bar{x}(t) \in A^a\}]^{\frac{1}{2}} + K^*\sqrt{a},$$
$$= \quad K^*\sqrt{a}.$$

Here the inequality follows by Lemma 1 and the equality by our choice of T, which implies in particular that the expectation on the right-hand side of the inequality is $O(a)$. Also, by comparing $x^t(\cdot)$ with a trajectory $x'(\cdot)$ in A and using the Gronwall inequality, we get

$$\|x^t(t+T) - x'(T)\| \leq K^*\|x^t(t) - x'(0)\| = K^*\|\bar{x}(t) - x'(0)\|.$$

Hence $\rho(x^t(t+T), A) \leq K^*\rho(\bar{x}(t), A)$. Since A is bounded, we also have $\rho(y, A)^2 \leq K^*(1 + \|y\|^2) \; \forall y$. Then

$$E[\rho(\bar{x}(t+T), A)^2 I\{\bar{x}(t) \in B(R)^c\}]^{\frac{1}{2}}$$
$$\leq \quad K^*\sqrt{a} + E[\rho(x^t(t+T), A)^2 I\{\bar{x}(t) \in B(R)^c\}]^{\frac{1}{2}}$$
$$\leq \quad K^*\sqrt{a} + K^* E[\rho(\bar{x}(t), A)^2 I\{\bar{x}(t) \in B(R)^c\}]^{\frac{1}{2}}$$
$$\leq \quad K^*\sqrt{a} + K^* E[(1 + \|\bar{x}(t)\|^2) I\{\bar{x}(t) \in B(R)^c\}]^{\frac{1}{2}}$$
$$\leq \quad K^*\sqrt{a}.$$

Here the first inequality follows from Lemma 1, the second and the third from the preceding observations, and the last from (9.2.8). This completes the proof. ∎

Theorem 3. *For a suitable constant $K > 0$,*

$$\limsup_{n \to \infty} E[\rho(x_n, A)^2]^{\frac{1}{2}} \leq K\sqrt{a}. \qquad (9.2.10)$$

Proof. Take $t = ma > 0$. By Lemma 1,

$$E[\|\bar{x}(t+T) - x^t(t+T)\|^2]^{\frac{1}{2}} \leq K'\sqrt{a}$$

for a suitable $K' > 0$. Then using (9.2.8), (9.2.9) and Lemma 2, one has:

$$E[\rho(x_{m+N}, A)^2]^{\frac{1}{2}}$$
$$= E[\rho(\bar{x}(t+T), A)^2]^{\frac{1}{2}}$$
$$\leq E[\rho(\bar{x}(t+T), A)^2 I\{\bar{x}(t) \notin B(R)\}]^{\frac{1}{2}}$$
$$\quad + E[\rho(\bar{x}(t+T), A)^2 I\{\bar{x}(t) \in B(R) - A^a\}]^{\frac{1}{2}}$$
$$\quad + E[\rho(\bar{x}(t+T), A)^2 I\{\bar{x}(t)) \in A^a\}]^{\frac{1}{2}}$$
$$\leq 2K^* \sqrt{a} + E[\rho(\bar{x}(t+T), A)^2 I\{\bar{x}(t) \in B(R) - A^a\}]^{\frac{1}{2}}$$
$$\leq 2K^* \sqrt{a} + E[\rho(x^t(t+T), A)^2 I\{\bar{x}(t) \in B(R) - A^a\}]^{\frac{1}{2}}$$
$$\quad + E[\|\bar{x}(t+T) - x^t(t+T)\|^2]^{\frac{1}{2}}$$
$$\leq (2K^* + K')\sqrt{a} + E[\rho(x^t(t+T), A)^2 I\{\bar{x}(t) \in B(R) - A^a\}]^{\frac{1}{2}}$$
$$\leq (2K^* + K')\sqrt{a} + \frac{1}{2} E[\rho(x^t(t), A)^2 I\{\bar{x}(t) \in B(R) - A^a\}]^{\frac{1}{2}}$$
$$= (2K^* + K')\sqrt{a} + \frac{1}{2} E[\rho(x_m, A)^2]^{\frac{1}{2}}.$$

Here the second inequality follows from Lemma 2 and the last one by (9.2.9). Iterating, one has

$$\limsup_{k \to \infty} E[\rho(x_{m+kN}, A)^2]^{\frac{1}{2}} \leq 2(2K^* + K')\sqrt{a}.$$

Repeating this for $m + 1, \ldots, m + N - 1$, in place of m, the claim follows. ∎

Let $\epsilon > 0$. By the foregoing and the Chebyshev inequality, we have

$$\limsup_{n \to \infty} P(\rho(x_n, A) > \epsilon) = O(a),$$

which captures the intuitive statement that '$\{x_n\}$ concentrate around A with a high probability as $n \to \infty$'.

9.3 Refinements

This section collects together the extensions to the constant stepsize scenario of various other results developed earlier for the decreasing stepsize case. In most cases, we only sketch the idea, as the basic philosophy is roughly similar to that for the decreasing stepsize case.

(i) *Stochastic recursive inclusions:* Consider

$$x_{n+1} = x_n + a[y_n + M_{n+1}], \quad n \geq 0,$$

with $y_n \in h(x_n)$ $\forall n$ for a set-valued map h satisfying the conditions stipulated at the beginning of Chapter 5. Let $\bar{x}(\cdot)$ denote the interpolated trajectory as in Chapter 5. For $T > 0$, let $\mathcal{S}_T$ denote the solution set of the differential inclusion

$$\dot{x}(t) \in h(x(t)), \ t \in [0, T]. \tag{9.3.1}$$

Under the 'linear growth' condition on $h(\cdot)$ stipulated in Chapter 5, viz., $\sup_{y \in h(x)} \|h(y)\| \leq K(1 + \|x\|)$, a straightforward application of the Gronwall inequality shows that the solutions to (9.3.1) remain uniformly bounded on finite time intervals for uniformly bounded initial conditions. For $z(\cdot) \in C([0, T]; \mathcal{R}^d)$, let

$$d(z(\cdot), \mathcal{S}_T) \overset{\text{def}}{=} \inf_{y(\cdot) \in \mathcal{S}_T} \sup_{t \in [0,T]} \|z(t) - y(t)\|.$$

For $t \geq 0$, suppose that:

(†)$E[\|\bar{x}(t)\|^2]$ remains bounded as $a \downarrow 0$. (This is a '*stability*' condition.)

Then the main result here is:

Theorem 4. $d(\bar{x}(\cdot)|_{[t',t'+T]}, \mathcal{S}_T) \overset{a \downarrow 0}{\to} 0$ *in law, uniformly in* $t' \geq 0$.

Proof. Fix $t' \in [n_0 a, (n_0 + 1)a), n_0 \geq 0$. Define $\tilde{x}(\cdot)$ by

$$\dot{\tilde{x}}(t) = y_{n_0+m}, \ t \in [(n_0 + m)a, (n_0 + m + 1)a) \cap [t', \infty),$$
$$0 \leq m < N, \tag{9.3.2}$$

with $\tilde{x}(t') = x_{n_0}$. Let $T = Na$ for simplicity. Then by familiar arguments,

$$E\left[\sup_{s \in [t', t'+T]} \|\bar{x}(s) - \tilde{x}(s)\|^2 \right]^{\frac{1}{2}} = O(\sqrt{a}). \tag{9.3.3}$$

By the 'stability' condition (†) mentioned above, the law of $\bar{x}(t')$ remains tight as $a \downarrow 0$. That is, $\tilde{x}(t')$, which coincides with $\bar{x}(t')$, remains tight in law. Also, for $t_1 < t_2$ in $[t', t' + T]$,

$$E[\|\tilde{x}(t_2) - \tilde{x}(t_1)\|^2] \leq |t_2 - t_1|^2 \sup_{n_0 \leq m \leq n_0+N} E[\|y_m\|^2]$$
$$\leq |t_2 - t_1|^2 K$$

for a suitable $K > 0$ independent of a. For the second inequality above, we have used the linear growth condition on $h(\cdot)$ along with the 'stability' condition (†) above. By the tightness criterion of Billingsley (1968), p. 95, it then follows that the laws of $\tilde{x}(t' + s), s \in [0, T]$, remain tight in $\mathcal{P}(C([0, T]; \mathcal{R}^d))$ as $a \to 0$ and t' varies over $[0, \infty)$. Thus along any sequences $a \approx a(k) \downarrow 0$, $\{t'(k)\} \subset [0, \infty)$, we can take a further subsequence, denoted $\{a(k), t'(k)\}$ again by abuse of notation, so that the $\tilde{x}(t'(k) + \cdot)$ converge in law. Denote $\tilde{x}(t'(k) + \cdot)$ by $\tilde{x}^k(\cdot)$ in order to make their dependence on $k, t'(k)$ explicit. By Skorohod's theorem (see Appendix C), there exist $C([0,T]; \mathcal{R}^d)$-valued random variables $\{\hat{x}^k(\cdot)\}$ such that $\hat{x}^k(\cdot), \tilde{x}^k(\cdot)$ agree in law separately for each k and $\{\hat{x}^k(\cdot)\}$ converge a.s. Now argue as in the proof of Theorem 1 of Chapter 5 to conclude that a.s., the limit thereof in $C([0,T]; \mathcal{R}^d)$ is in $\mathcal{S}_T$, i.e.,

$$d(\hat{x}^k(\cdot), \mathcal{S}_T) \to 0 \text{ a.s.}$$

Thus

$$d(\tilde{x}^k(\cdot), \mathcal{S}_T) \to 0 \text{ in law.}$$

In view of (9.3.3), we then have

$$d(\tilde{x}^k(t'(k) + \cdot), \mathcal{S}_T) \to 0 \text{ in law,}$$

where the superscript k renders explicit the dependence on the stepsize $a(k)$. The claim follows. ∎

If the differential inclusion (9.3.1) has a globally asymptotically stable compact attractor A, we may use this in place of Lemma 1 to derive a variation of Theorem 3 for the present set-up.

(ii) *Avoidance of traps:* This is rather easy in the constant stepsize set-up. Suppose (9.1.2) has compact attractors $A_1, \ldots, A_M$ with respective domains of attraction $D_1, \ldots, D_M$. Suppose that the set $U \stackrel{\text{def}}{=}$ the complement of $\cup_{i=1}^M D_i$, has the following property: For any $\epsilon > 0$, there exists an open neighbourhood U^ϵ of U and a constant $\hat{C} > 0$ such that for $A = \cup_i A_i$,

$$\limsup_{n \to \infty} E[\rho(x_n, A^a)^2 I\{x_n \in U^\epsilon\}] \leq \hat{C}a. \tag{9.3.4}$$

Intuitively, this says that the 'bad' part of the state space has low probability in the long run. One way (9.3.4) can arise is if U has zero Lebesgue measure and the laws of the x_n have uniformly bounded densities w.r.t. the Lebesgue measure on $\mathcal{R}^d$. One then chooses ϵ small enough that (9.3.4) is met. Under (9.3.4), we can modify the calculation in the proof of Theorem 3 as

$$E[\rho(x_{m+N}, A)^2]^{\frac{1}{2}}$$
$$\leq E[\rho(\bar{x}(t+T), A)^2 I\{\bar{x}(t) \notin B(R)\}]^{\frac{1}{2}}$$
$$+ E[\rho(\bar{x}(t+T), A)^2 I\{\bar{x}(t) \in U^\epsilon\}]^{\frac{1}{2}}$$
$$+ E[\rho(\bar{x}(t+T), A)^2 I\{\bar{x}(t) \in B(R) \cap (U^\epsilon \cup A^a)^c\}]^{\frac{1}{2}}$$
$$+ E[\rho(\bar{x}(t+T), A)I\{\bar{x}(t)) \in A^a\}]^{\frac{1}{2}}$$

Choosing T appropriately as before, argue as before to obtain (9.2.10), using (9.3.4) in addition to take care of the second term on the right.

(iii) *Stability:* We now sketch how the first stability criterion described in section 3.2 can be extended to the constant stepsize framework. The claim will be that $\sup_n E[||x_n||^2]$ remains bounded for a *sufficiently small* stepsize a.

As in section 3.2, let $h_c(x) \stackrel{\text{def}}{=} h(cx)/c \; \forall x, 1 \le c < \infty$, and $h_\infty(\cdot) \stackrel{\text{def}}{=}$ limit in $C(\mathcal{R}^d)$ of $h_c(\cdot)$ as $c \uparrow \infty$, assumed to exist. Let assumption (A5) there hold, i.e., the o.d.e.

$$\dot{x}_\infty(t) = h_\infty(x_\infty(t))$$

have the origin as the unique globally asymptotically stable equilibrium point. Let $x_c(\cdot)$ denote a solution to the o.d.e.

$$\dot{x}_c(t) = h_c(x_c(t)), \tag{9.3.5}$$

for $c \ge 1$. Then by Corollary 3 of Chapter 3, there exists $c_0 > 0, T > 0$, such that whenever $||x_c(0)|| = 1$, one has

$$||x_c(t)|| < \frac{1}{4} \text{ for } t \in [T, T+1] \; \forall \; c \ge c_0. \tag{9.3.6}$$

We take $T = Na$ for some $N \ge 1$ without loss of generality, which specifies N as a function of T and a. Let $T_n = nNa, n \ge 0$. By analogy with what we did for the decreasing stepsize case in Chapter 3, define $\hat{x}(\cdot)$ as follows. On $[T_n, T_{n+1}]$, define

$$\tilde{x}^n((nN+k)a) \stackrel{\text{def}}{=} \frac{x_{nN+k}}{||x_{nN}|| \vee 1}, \; 0 \le k \le N,$$

with linear interpolation. Define

$$\hat{x}(t) \stackrel{\text{def}}{=} \lim_{s \downarrow t} \tilde{x}^n(s), \;\; t \in [T_n, T_{n+1}).$$

Let $\hat{x}(T_{n+1}-) \stackrel{\text{def}}{=} \lim_{t \uparrow T_{n+1}} \hat{x}(t)$ for $n \ge 0$. Then $\hat{x}(\cdot)$ is piecewise linear and continuous except possibly at the T_n, where it will be right continuous with its left limit well defined. Let $x^n(\cdot)$ denote a solution to (9.3.5) on $[T_n, T_{n+1}]$ with $c = ||x_{nN}|| \vee 1$ and $x^n(T_n) = \hat{x}(T_n)$ for $n \ge 0$. By the arguments of section 9.2,

$$E[\sup_{t \in [T_n, T_{n+1})} ||\hat{x}(t) - x^n(t)||^2]^{\frac{1}{2}} \le C_1 \sqrt{a}, \; \forall n, \tag{9.3.7}$$

for a suitable constant $C_1 > 0$ independent of a. As before, let $\mathcal{F}_n = \sigma(x_i, M_i, i \le n), \; n \ge 0$.

Lemma 5. *For $n \geq 0$ and a suitable constant $C_2 > 0$ depending on T,*

$$\sup_{0 \leq k \leq N} E[\||x_{nN+k}\||^2|\mathcal{F}_{nN}]^{\frac{1}{2}} \leq C_2(1 + \||x_{nN}\||) \quad a.s. \; \forall \, n \geq 0.$$

Proof. Recall that

$$E[\||M_{n+1}\||^2|\mathcal{F}_n] \leq K_3(1 + \||x_n\||^2) \; \forall n, \tag{9.3.8}$$

for some $K_3 > 0$ (cf. assumption (A3) of Chapter 2). By (9.1.1), for $n \geq 0, 0 \leq k \leq N$,

$$E[\||x_{nN+k+1}\||^2|\mathcal{F}_{nN}]^{\frac{1}{2}}$$
$$\leq \quad E[\||x_{nN+k}\||^2|\mathcal{F}_{nN}]^{\frac{1}{2}} + aK'(1 + E[\||x_{nN+k}\||^2|\mathcal{F}_{nN}]^{\frac{1}{2}}),$$

for a suitable $K' > 0$, where we use (9.3.8) and the linear growth condition on h. The claim follows by iterating this inequality. ∎

Theorem 6. *For sufficiently small $a > 0$, $\sup_n E[\||x_n\||^2] < \infty$.*

Proof. Let c_0, T be such that (9.3.6) holds and pick $\sqrt{a} \leq (2C_1)^{-1}$ for C_1 as in (9.3.7). For $n \geq 0$, $0 \leq k \leq N$, we have

$$E[\||x_{(n+1)N}\||^2|\mathcal{F}_{nN}]^{\frac{1}{2}}$$
$$= \quad E[\||\tilde{x}_{(n+1)N}\||^2|\mathcal{F}_{nN}]^{\frac{1}{2}}(\||x_{nN}\|| \vee 1)$$
$$\leq \quad E[\||\tilde{x}_{(n+1)N} - x^n(T_{n+1})\||^2|\mathcal{F}_{nN}]^{\frac{1}{2}}(\||x_{nN}\|| \vee 1)$$
$$\quad + E[\||x^n(T_{n+1})\||^2|\mathcal{F}_{nN}]^{\frac{1}{2}}(\||x_{nN}\|| \vee 1)$$
$$\leq \quad \frac{1}{2}\||x_{nN}\|| \vee 1$$
$$\quad + E[\||x^n(T_{n+1})\||^2 I\{\||x_{nN}\|| \geq c_0\}|\mathcal{F}_{nN}]^{\frac{1}{2}}(\||x_{nN}\|| \vee 1)$$
$$\quad + E[\||x^n(T_{n+1})\||^2 I\{\||x_{nN}\|| < c_0\}|\mathcal{F}_{nN}]^{\frac{1}{2}}(\||x_{nN}\|| \vee 1)$$
$$\leq \quad \frac{1}{2}\||x_{nN}\|| + \frac{1}{4}\||x_{nN}\|| + \bar{C}$$
$$= \quad \frac{3}{4}\||x_{nN}\|| + \bar{C},$$

where the second inequality follows by (9.3.7) and our choice of a, and the third inequality follows from (9.3.6), (9.3.7) and Lemma 5 above, with $\bar{C} \overset{\text{def}}{=} C_2(1 + c_0) + 1$. Thus

$$E[\||x_{(n+1)N}\||^2]^{\frac{1}{2}} \leq \frac{3}{4}E[\||x_{nN}\||^2]^{\frac{1}{2}} + \bar{C}.$$

By iterating this inequality, we have $\sup_n E[\||x_{nN}\||^2] < \infty$, whence the claim follows by Lemma 5. ∎

(iv) *Two timescales:* By analogy with section 6.1, consider the coupled iterations

$$x_{n+1} = x_n + a[h(x_n, y_n) + M_{n+1}^1], \qquad (9.3.9)$$
$$y_{n+1} = y_n + b[g(x_n, y_n) + M_{n+1}^2], \qquad (9.3.10)$$

for $0 < b << a$, where h, g are Lipschitz and $\{M_{n+1}^i\}, i = 1, 2$, are martingale difference sequences satisfying

$$E[||M_{n+1}^i||^2 | M_m^j, x_m, y_m, m \le n, j = 1, 2]$$
$$\le C(1 + ||x_n||^2 + ||y_n||^2),$$

for $i = 1, 2$. Assume (9.2.1) and (9.2.2) for $\{x_n\}$ along with their counterparts for $\{y_n\}$. We also assume that the o.d.e.

$$\dot{x}(t) = h(x(t), y) \qquad (9.3.11)$$

has a unique globally asymptotically stable equilibrium $\lambda(y)$ for each y and a Lipschitz function $\lambda(\cdot)$, and that the o.d.e.

$$\dot{y}(t) = h(\lambda(y(t)), y(t)) \qquad (9.3.12)$$

has a unique globally asymptotically stable equilibrium y^*. Then the arguments of section 6.1 may be combined with the arguments of section 9.2 to conclude that

$$\limsup_{n \to \infty} E[||x_n - \lambda(y^*)||^2 + ||y_n - y^*||^2] = O(a) + O(\frac{b}{a}). \qquad (9.3.13)$$

Specifically, consider first the timescale corresponding to the stepsize a and consider the interpolated trajectory $\bar{x}(\cdot)$ on $[na, na + T]$ for $T \stackrel{\text{def}}{=} Na > 0$, for some $N \ge 1$ and $n \ge 0$. Let $x^n(\cdot)$ denote the solution of the o.d.e. (9.3.11) on $[na, na + T]$ for $y = y_n$ and $x^n(na) = x_n$. Then arguing as for Lemma 1 above,

$$E[\sup_{t \in [na, na+T]} ||\bar{x}(t) - x^n(t)||^2] = O(a) + O(\frac{b}{a}).$$

Here we use the easily established fact that

$$\sup_{0 \le k \le N} E[||y_{n+k} - y_n||^2] = O(\frac{b}{a}),$$

and thus the approximation $y_{n+k} \approx y_n, 0 \le k \le N$, contributes only another $O(\frac{b}{a})$ error. Given our hypotheses on the asymptotic behaviour of (9.3.11), it follows that

$$\limsup_{n \to \infty} E[||x_n - \lambda(y_n)||^2] = O(a) + O(\frac{b}{a}).$$

Next, consider the timescale corresponding to b. Let $T' \overset{\text{def}}{=} Mb > 0$ for some $M \geq 1$. Consider the interpolated trajectory $\bar{y}(\cdot)$ on $[nb, nb + T']$ defined by $\bar{y}(mb) \overset{\text{def}}{=} y_m \; \forall m$, with linear interpolation. Let $y^n(\cdot)$ denote the solution to (9.3.12) on $[nb, nb + T']$ with $y^n(nb) = y_n$ for $n \geq 0$. Then argue as in the proof of Lemma 1 to conclude that

$$\limsup_{n \to \infty} E[\sup_{t \in [na, na+T']} \|\bar{y}(t) - y^n(t)\|^2] = O(a) + O(\frac{b}{a}).$$

The only difference from the argument leading to Lemma 1 is an additional error term due to the approximation $x_n \approx \lambda(y_n)$, which is $O(a) + O(\frac{b}{a})$ as observed above. (This, in fact, gives the $O(a) + O(\frac{b}{a})$ on the right-hand side instead of $O(b)$.) Given our hypotheses on (9.3.12), this implies that

$$\limsup_{n \to \infty} E[\|y_n - y^*\|^2] = O(a) + O(\frac{b}{a}),$$

which in turn will also yield (in view of the Lipschitz continuity of $\lambda(\cdot)$)

$$\limsup_{n \to \infty} E[\|x_n - \lambda(y^*)\|^2] = O(a) + O(\frac{b}{a}).$$

(v) *Averaging the 'natural' timescale:* Now consider

$$x_{n+1} = x_n + a[h(x_n, Y_n) + M_{n+1}], \; n \geq 0, \tag{9.3.14}$$

where $\{Y_n\}$ is as in sections 6.2 and 6.3. That is, it is a process taking values in a complete separable metric space S such that for any Borel set $A \subset S$,

$$P(Y_{n+1} \in A | Y_m, Z_m, x_m, m \leq n) = \int_A p(dy | Y_n, Z_n, x_n), \; n \geq 0.$$

Here $\{Z_n\}$ takes values in a compact metric space U and $p(dy|\cdot, \cdot, \cdot)$ is a continuous 'controlled' transition probability kernel. Mimicking the developments of Chapter 6, define

$$\mu(t) \overset{\text{def}}{=} \delta_{(Y_n, Z_n)}, \; t \in [na, (n+1)a), \; n \geq 0,$$

where $\delta_{(y,z)}$ is the Dirac measure at (y, z). For $s \geq 0$, let $x^s(t), t \geq s$, be the solution to the o.d.e.

$$\dot{x}^s(t) = \tilde{h}(x^s(t), \mu(t)) \overset{\text{def}}{=} \int h(x^s(t), \cdot) d\mu(t). \tag{9.3.15}$$

Define $\bar{x}(\cdot)$ as at the start of section 9.2. Then by familiar arguments,

$$E[\sup_{t \in [s, s+T]} \|\bar{x}(t) - x^s(t)\|^2] = O(a). \tag{9.3.16}$$

Assume the 'stability condition' (†) above. It then follows by familiar arguments that, as $a \downarrow 0$, the laws of $x^s(\cdot)|_{[s,s+T]}, s \geq 0$, remain tight as probability measures on $C([0,T]; \mathcal{R}^d)$. Suppose the laws of $\mu(s + \cdot)$ remain tight as well. Then every sequence $a(n) \downarrow 0, s = s(a(n)) \in [0, \infty)$, has a further subsequence, denoted by $\{a(n), s(a(n))\}$ again by abuse of terminology, such that the corresponding processes $(x^s(\cdot), \mu(s + \cdot))$ converge in law. Invoking Skorohod's theorem, we may suppose that this convergence is a.s. We shall now need a counterpart of Lemma 6 of Chapter 6. Let $\{f_i\}$ be as in the proof of Lemma 6, Chapter 6, and define $\{\xi_n^i\}$, $\tau(n,t)$ as therein, with $a(m) \equiv a$. It is then easily verified that almost surely as $a \downarrow 0$ along $\{a(\ell)\}$, we have

$$E[(\sum_{m=n}^{\tau(n,t)} a(f(Y_{m+1}) - \int f(y)p(dy|Y_m, Z_m, x_m))^2]$$

$$= E[\sum_{m=n}^{\tau(n,t)} a^2(f(Y_{m+1}) - \int f(y)p(dy|Y_m, Z_m, x_m))^2]$$

$$= O(a) \overset{a\downarrow 0}{\rightarrow} 0.$$

As for Lemma 6, Chapter 6, this leads to the following: almost surely, for any limit point $(\tilde{x}(\cdot), \tilde{\mu}(\cdot))$ of $(x^s(\cdot), \mu(s + \cdot))$ as $a \downarrow 0$ along $\{a(\ell)\}$,

$$\int_0^t (f_i(y) - \int f_i(w)p(dw|y, z, \tilde{x}(s))\tilde{\mu}(s)(dydz))ds = 0$$

$\forall i \geq 1, t \geq 0$. Argue as in the proof of the lemma to conclude that $\tilde{\mu}(t) \in D(\tilde{x}(t))$ $\forall t$. It then follows that a.s., $x^s(\cdot)|_{[s,s+T]}$ converges to $\mathcal{G}_T \overset{\text{def}}{=}$ the set of trajectories of the differential inclusion

$$\dot{x}(t) \in \hat{h}(x(t)), \quad s \leq t \leq s + T, \tag{9.3.17}$$

for a set-valued map $\hat{h}$ defined as in section 6.3. Set

$$\bar{d}(z(\cdot), \mathcal{G}_T) \overset{\text{def}}{=} \inf_{y(\cdot) \in \mathcal{G}_T} \sup_{t \in [0,T]} ||z(t) - y(t)||, \quad z(\cdot) \in C([0,T]; \mathcal{R}^d).$$

Then we can argue as in extension (i) above to conclude:

Theorem 7. *For any $T > 0$, $\bar{d}(\bar{x}(\cdot)|_{[s,s+T]}, \mathcal{G}_T) \overset{a\downarrow 0}{\rightarrow} 0$ in law uniformly in $s \geq 0$.*

The asymptotic behaviour of the algorithm as $n \to \infty$ may then be inferred from the asymptotic behaviour of trajectories in $\mathcal{G}_T$ as in Chapter 5. As in Chapter 6, consider the special case when there is no 'external control process' $\{Z_n\}$ in the picture and in addition, for

$x_n \equiv x \in \mathcal{R}^d$, the process $\{Y_n\}$ is an ergodic Markov process with the unique stationary distribution $\nu(x)$. Then (9.3.17) reduces to

$$\dot{x}(t) = \bar{h}(x(t), \nu(x(t))).$$

(vi) *Asynchronous implementations:* This extension proceeds along lines similar to that for decreasing stepsize, with the corresponding claims adapted as in section 9.2. Thus, for example, for the case with no inter-processor delays, we conclude that for $t \geq 0$,

$$E[\sup_{s \in [t,t+T]} \|\bar{x}(s) - \tilde{x}^t(s)\|^2] = O(a),$$

where $\tilde{x}(s), s \in [t, t+T]$, is a trajectory of

$$\dot{\tilde{x}}(s) = \Lambda(s)h(\tilde{x}(s)), \quad \tilde{x}(t) = \bar{x}(t),$$

for a $\Lambda(\cdot)$ as in Theorem 2 of Chapter 7. The delays simply contribute another $O(a)$ error term and thus do not affect the conclusions.

One can use this information to 'rig' the stepsizes so as to get the desired limiting o.d.e. when a common clock is available. This is along the lines of the concluding remarks of section 7.4. Thus, for example, suppose the components are updated one at a time according to an ergodic Markov chain $\{Y_n\}$ on their index set. That is, at time n, the Y_nth component is being updated. Suppose the chain has a stationary probability vector $[\pi_1, \ldots, \pi_d]$. Then by Corollary 8 of Chapter 6, $\Lambda(t) \equiv \mathrm{diag}(\pi_1, \ldots, \pi_d)$. Thus if we use the stepsize a/π_i for the ith component, we get the limiting o.d.e. $\dot{x}(t) = h(x(t))$ as desired. In practice, we may use $a/\eta_n(i)$ instead, where $\eta_n(i)$ is an empirical estimate of π_i obtained by suitable averaging on a faster timescale, so that it tracks π_i. As in Chapter 7, this latter arrangement also extends in a natural manner to the more general case when $\{Y_n\}$ is 'controlled Markov' as in Chapter 6.

(vii) *Limit theorems:* For $T = Na > 0$ as above, let $x^s(\cdot), s = na$ (say), denote the solution of (9.1.2) on $[s, s+T]$ with $x^s(s) = x_n$ for $n \geq 0$. We fix T and vary a, with $N = \lceil \frac{T}{a} \rceil$, $s = na$. Define $z^n(t), t \in [na, na+T]$, by

$$z^n((n+k)a) \stackrel{\text{def}}{=} \frac{1}{\sqrt{a}}(x_{n+k} - x^s((n+k)a)), \quad 0 \leq k \leq N,$$

with linear interpolation. Then arguing as in sections 8.2 and 8.3 (with the additional hypotheses therein), we conclude that the limits in law as $n \to \infty$ of the laws of $\{z^n(\cdot)\}$, viewed as $C([0, T]; \mathcal{R}^d)$-valued random variables, are the laws of a random process on $[0, T]$ of the form

$$z^*(t) = \int_0^t \nabla h(x^{*s}(s))z^*(s)ds + \int_0^t D(x^{*s}(s))dB(s), \quad t \in [0, T],$$

where $D(\cdot)$ is as in Chapter 8, $x^{*s}(\cdot)$ is a solution of (9.1.2), and $B(\cdot)$ is a standard Brownian motion in $\mathcal{R}^d$. If we let $s \uparrow \infty$ as well and (9.1.2) has a globally asymptotically stable compact attractor A, $x^{*s}(\cdot)$ will concentrate with high probability in a neighbourhood of the set $\{x(\cdot) \in C([0,T];\mathcal{R}^d) : x(\cdot) \text{ satisfies (9.1.2) with } x(0) \in A\}$. Furthermore, in the special case when (9.1.2) has a unique globally asymptotically stable equilibrium x_{eq}, $x^*(\cdot) \equiv x_{eq}$, $z^*(\cdot)$ is a Gauss–Markov process, and we recover the central limit theorem for $\{x_n\}$.

10

Applications

10.1 Introduction

This chapter is an overview of several applications of stochastic approximation in broad strokes. These examples are far from exhaustive and are meant to give only a flavour of the immense potentialities of the basic scheme. In each case, only the general ideas are sketched. The details are relatively routine in many cases. In some cases where they are not, pointers to the relevant literature are provided.

The applications have been broadly classified into the following three categories:

(i) *Stochastic gradient schemes:* These are stochastic approximation versions of classical gradient ascent or descent for optimizing some performance measure.

(ii) *Stochastic fixed point iterations:* These are stochastic approximation versions of classical fixed point iterations $x_{n+1} = f(x_n)$ for solving the fixed point equation $x = f(x)$.

(iii) *Collective phenomena:* This is the broad category consisting of diverse models of interacting autonomous agents arising in engineering or economics.

In addition we have some miscellaneous instances that don't quite fit any of the categories above. In several cases, we shall only look at the limiting o.d.e. This is because the asymptotic behaviour of the actual stochastic iteration can be easily read off from this o.d.e. in view of the theory developed so far in this book.

117

10.2 Stochastic gradient schemes

Stochastic gradient schemes are iterations of the type

$$x_{n+1} = x_n + a(n)[-\nabla f(x_n) + M_{n+1}],$$

where $f(\cdot)$ is the continuously differentiable function we are seeking to minimize and the expression in square brackets represents a noisy measurement of the gradient. (We drop the minus sign on the right-hand side when the goal is *maximization*.) Typically $\lim_{\|x\| \to \infty} f(x) = \infty$, ensuring the existence of a global minimum for f. The limiting o.d.e. then is

$$\dot{x}(t) = -\nabla f(x(t)), \tag{10.2.1}$$

for which f itself serves as a 'Liapunov function':

$$\frac{d}{dt} f(x(t)) = -\|\nabla f(x(t))\|^2 \leq 0,$$

with a strict inequality when $\nabla f(x(t)) \neq 0$. Let $H \stackrel{\text{def}}{=} \{x : \nabla f(x) = 0\}$ denote the set of equilibrium points for this o.d.e. Recall the definition of an ω-limit set from Appendix B.

Lemma 1. *The only possible invariant sets that can occur as ω-limit sets for (10.2.1) are the subsets of H.*

Proof. If the statement is not true, there exists a trajectory $x(\cdot)$ of (10.2.1) such that its ω-limit set contains a non-constant trajectory $\tilde{x}(\cdot)$. By the foregoing observations, $f(\tilde{x}(t))$ must be monotonically decreasing. Let $t > s$, implying $f(\tilde{x}(t)) < f(\tilde{x}(s))$. But by the definition of an ω-limit set, we can find $t_1 < s_1 < t_2 < s_2 < \cdots$ such that $x(t_n) \to \tilde{x}(t)$ and $x(s_n) \to \tilde{x}(s)$. It follows that for sufficiently large n, $f(x(t_n)) < f(x(s_n))$. This contradicts the fact that $f(x(\cdot))$ is monotonically decreasing, proving the claim. ∎

Suppose H is a discrete set. Assume f to be twice continuously differentiable. Then ∇f is continuously differentiable and its Jacobian matrix, i.e., the Hessian matrix of f, is positive definite at $x \in H$ if and only if x is a local minimum. Thus linearizing the o.d.e. around any $x \in H$, we see that the local minima are the stable equilibria of the o.d.e. and the 'avoidance of traps' results in Chapter 3 tell us that $\{x_n\}$ will converge a.s. to a local minimum under reasonable conditions. (Strictly speaking, one should allow the situation when both the first and the second derivatives vanish at a point in H. We ignore this scenario as it is non-generic.)

The assumption that H is discrete also seems reasonable in view of the result from Morse theory that f with isolated critical points are dense in $C(\mathcal{R}^d)$ (see, e.g., Chapter 2 of Matsumoto, 2002). But this has to be taken with a pinch of salt. In many stochastic-approximation-based parameter tuning schemes in engineering, non-isolated equilibria can arise due to overparametrization.

There are several variations on the basic scheme, mostly due to the unavailability of even a noisy measurement of the gradient assumed in the foregoing. In many cases, one needs to approximately evaluate the gradient. Thus we may replace the scheme above by

$$x_{n+1} = x_n + a(n)[-\nabla f(x_n) + M_{n+1} + \eta(n)],$$

where $\{\eta(n)\}$ is the additional 'error' in gradient estimation. Suppose one has

$$\sup_n \|\eta(n)\| < \epsilon_0$$

for some small $\epsilon_0 > 0$. Then by Theorem 6 of Chapter 5, the iterates converge a.s. to a small neighbourhood of some point in H. (The smaller the ϵ_0 we are able to pick, the better the prospects.) This result may be further refined by adapting the 'avoidance of traps' argument of Chapter 4 to argue that the convergence is in fact to a neighbourhood of some local minimum.

The simplest such scheme, going back to Kiefer and Wolfowitz (1952), uses a finite difference approximation. Let $x_n \overset{\text{def}}{=} [x_n(1), \ldots, x_n(d)]^{\mathrm{T}}$ and similarly, $M_n \overset{\text{def}}{=} [M_n(1), \ldots, M_n(d)]^{\mathrm{T}}$. Let $e_i \overset{\text{def}}{=}$ denote the unit vector in the ith coordinate direction for $1 \leq i \leq d$ and $\delta > 0$ a small positive scalar. The algorithm is

$$x_{n+1}(i) = x_n(i) + a(n)[-\left(\frac{f(x_n + \delta e_i) - f(x_n - \delta e_i)}{2\delta}\right) + M_{n+1}(i)]$$

for $1 \leq i \leq d$, $n \geq 0$. (M_{n+1} here collects together the net 'noise' in all the function evaluations involved.) By Taylor's theorem, the error in replacing the gradient with its finite difference approximation as above is $O(\delta \|\nabla^2 f(x_n)\|)$, where $\nabla^2 f$ denotes the Hessian matrix of f. If this error is small, the foregoing analysis applies. (A further possibility is to slowly reduce δ to zero, whence the accuracy of the approximation improves. But usually the division by δ would also feature in the martingale difference term M_{n+1} above and there is a clear trade-off between improvement of the mean error due to finite difference approximation alone and increased fluctuation and numerical problems caused by the small denominator.)

Note that this scheme requires $2d$ function evaluations. If one uses 'one-sided differences' to replace the algorithm above by

$$x_{n+1}(i) = x_n(i) + a(n)[-\left(\frac{f(x_n + \delta e_i) - f(x_n)}{\delta}\right) + M_{n+1}(i)],$$

the number of function evaluations is reduced to $d+1$, which may still be high for many applications. A remarkable development in this context is the *simultaneous perturbation stochastic approximation* (SPSA) due to Spall (1992). Let $\{\Delta_n(i), 1 \le i \le d, n \ge 0\}$ be i.i.d. random variables such that

(i) $\Delta_n \overset{\text{def}}{=} [\Delta_n(1), \ldots, \Delta_n(d)]^{\mathrm{T}}$ is independent of $M_{i+1}, x_i, i \le n; \Delta_j, j < n$, for each $n \ge 0$, and

(ii) $P(\Delta_m(i) = 1) = P(\Delta_m(i) = -1) = \frac{1}{2}$.

Considering the one-sided scheme for simplicity, we replace the algorithm above by

$$x_{n+1}(i) = x_n(i) + a(n)[-\left(\frac{f(x_n + \delta\Delta_n) - f(x_n)}{\delta\Delta_n(i)}\right) + M_{n+1}(i)],$$

for $n \ge 0$. Note that by Taylor's theorem, for each i,

$$\left(\frac{f(x_n + \delta\Delta_n) - f(x_n)}{\delta\Delta_n(i)}\right) \approx \frac{\partial f}{\partial x_i}(x_n) + \sum_{j \ne i} \frac{\partial f}{\partial x_j}(x_n)\frac{\Delta_n(j)}{\Delta_n(i)}.$$

The expected value of the second term on the right is zero. Hence it acts as just another noise term like M_{n+1}, averaging out to zero in the limit. Thus this is a valid approximate gradient scheme which requires only two function evaluations. A two-sided counterpart can be formulated similarly. Yet another variation which requires a single function evaluation is

$$x_{n+1}(i) = x_n(i) + a(n)[-\left(\frac{f(x_n + \delta\Delta_n)}{\delta\Delta_n(i)}\right) + M_{n+1}(i)],$$

which uses the fact that

$$\left(\frac{f(x_n + \delta\Delta_n)}{\delta\Delta_n(i)}\right) \approx \frac{f(x_n)}{\delta\Delta_n(i)} + \frac{\partial f}{\partial x_i}(x_n) + \sum_{j \ne i} \frac{\partial f}{\partial x_j}(x_n)\frac{\Delta_n(j)}{\Delta_n(i)}.$$

Both the first and the third term on the right average out to zero in the limit, though the small δ in the denominator of the former degrades the performance. See Chapter 7 of Spall (2003) for a comparison of the two alternatives from a practical perspective. In general, one can use more general $\Delta_n(i)$ as long as they are i.i.d. zero mean and $\Delta_n(i), \Delta_n(i)^{-1}$ satisfy suitable moment conditions – see Spall (2003). Bhatnagar et al. (2003) instead use cleverly chosen deterministic sequences to achieve the same effect, with some computational advantages.

Another scheme which works with a single function evaluation at a time is that of Katkovnik and Kulchitsky (1972). The idea is as follows: Suppose we replace ∇f by its approximation

$$Df_\sigma(x) \overset{\text{def}}{=} \int G_\sigma(x - y)\nabla f(y)dy,$$

where $G_\sigma(\cdot)$ is the Gaussian density with mean zero and variance σ^2 and the integral is componentwise. This is a good approximation to ∇f for small values of σ^2. Integrating by parts, we have

$$Df_\sigma(x) = \int \nabla G_\sigma(x - y) f(y) dy,$$

where the right-hand side can be cast as another (scaled) Gaussian expectation. Thus it can be approximated by a Monte Carlo technique which may be done either separately in a batch mode, or on a faster timescale as suggested in section 6.1. This scheme has the problem of numerical instability due to the presence of a small term σ^2 that appears in the denominator of the actual computation and may need smoothing and/or truncation to improve its behaviour. See section 7.6 of Rubinstein (1981) for the general theoretical framework and variations on this idea.

Note that the scheme

$$x_{n+1} = x_n + a(n)[h(x_n) + M_{n+1}], \ n \ge 0,$$

will achieve the original objective of minimizing f for any $h(\cdot)$ that satisfies

$$\langle \nabla f(x), h(x) \rangle < 0 \ \ \forall \ x \notin H.$$

We shall call such schemes *gradient-like*. One important instance of these is a scheme due to Fabian (1960). In this, $h(x) = -\text{sgn}(\nabla f(x))$, where the ith component of the right-hand side is simply $+1$ or -1 depending on whether the ith component of $\nabla f(x)$ is < 0 or > 0, and is zero if the latter is zero. Thus

$$\langle h(x), \nabla f(x) \rangle = - \sum_i |\frac{\partial f}{\partial x_i}(x)| < 0 \ \forall x \notin H.$$

This scheme typically has more graceful, but slower behaviour away from H. Since the $\text{sgn}(\cdot)$ function defined above is discontinuous, one has to invoke the theory of stochastic recursive inclusions to analyze it as described in section 5.3, under the heading *Discontinuous dynamics*. That is, one considers the limiting differential inclusion

$$\dot{x}(t) \in \bar{h}(x(t)),$$

where the ith component of the set-valued map $\bar{h}(x)$ is $\{+1\}$ or $\{-1\}$ depending on whether the ith component of $h(x)$ is > 0 or < 0, and is $[-1, 1]$ if it is zero. In practical terms, the discontinuity leads to some oscillatory behaviour when a particular component is near zero, which can be 'smoothed' out by taking a smooth approximation to $\text{sgn}(\cdot)$ near its discontinuity.

In the optimization literature, there are improvements on basic gradient descent such as the conjugate gradient and Newton/quasi-Newton methods. The stochastic approximation variants of these have also been investigated, see, e.g., Anbar (1978), Ruppert (1985) and Ruszczynski and Syski (1983).

A related situation arises when one is seeking a saddle point of a function $f(\cdot,\cdot) : A \times B \subset \mathcal{R}^n \times \mathcal{R}^n \to \mathcal{R}$, i.e., a point (x^*, y^*) such that

$$\min_x \max_y f(x,y) = \max_y \min_x f(x,y) = f(x^*, y^*).$$

This is known to exist, e.g., when A, B are compact convex and $f(\cdot, y)$ (resp. $f(x, \cdot)$) is convex (resp. concave) for each fixed y (resp. x). Given noisy measurements of the corresponding partial derivatives, one may then perform (say) stochastic gradient descent $\{x_n\}$ w.r.t. the x-variable on the fast timescale and stochastic gradient ascent $\{y_n\}$ w.r.t. the y-variable on the slow timescale. By our arguments of Chapter 6, $\{x_n\}$ will asymptotically track $g(y_n) \overset{\text{def}}{=} \operatorname{argmin}(f(\cdot, y_n))$, implying that $\{y_n\}$ tracks the o.d.e.

$$\dot{y}(t) = \nabla_y f(z,y)|_{z=g(y)}.$$

Here ∇_y is the gradient in the y-variable. Assuming the uniqueness of the saddle point, $y(\cdot)$ and therefore $\{y_n\}$ will then converge (a.s. in the latter case) to y^* under reasonable conditions if we are able to rewrite the o.d.e. above as

$$\dot{y}(t) = \nabla_y \min_x f(x,y),$$

i.e., to claim that $\nabla_y \min_x f(x,y) = \nabla_y f(z,y)|_{z=g(y)}$. This is true under very stringent conditions, and is called the 'envelope theorem' in mathematical economics. Some recent extensions thereof (Milgrom and Segal, 2002; see also Bardi and Capuzzo-Dolcetta, 1997, pp. 42–46) sometimes allow one to extend this reasoning to more general circumstances. See Borkar (2005) for one such situation.

An important domain of related activity is that of *simulation-based optimization*, wherein one seeks to maximize a performance measure and either this measure or its gradient is to be estimated from a simulation. There are several important strands of research in this area and we shall very briefly describe a few. To start with, consider the problem of maximizing over a real parameter θ a performance measure $J(\theta) \overset{\text{def}}{=} E_\theta[f(X)]$, where f is a nice (say, continuous) function $\mathcal{R} \to \mathcal{R}$ and $E_\theta[\,\cdot\,]$ denotes the expectation of the real random variable X whose distribution function is F_θ. The idea is to update the guesses $\{\theta(n)\}$ for the optimal θ based on simulated values of pseudo-random variables $\{X_n\}$ generated such that the distribution function of X_n is $F_{\theta(n)}$ for each n. Typically one generates a random variable X with a prescribed continuous and strictly increasing distribution function F by taking $X = F^{-1}(U)$, where U is uniformly distributed on $[0,1]$. A slightly messier expression works when F is either discontinuous or not strictly increasing. More generally, one has $X = \Psi(U)$ for U as above and a suitable Ψ. Thus we may suppose that $f(X_n) = \Phi(U_n, \theta(n))$ for $\{U_n\}$ i.i.d. uniform on $[0,1]$ and some suitable Φ. Suppose $\Phi(u, \cdot)$ is continuously differentiable and the interchange of expectation and differentiation in

$$\frac{d}{d\theta}E[\Phi(U,\theta)] = E[\frac{\partial}{\partial\theta}\Phi(U,\theta)]$$

is justified. Then a natural scheme would be the stochastic approximation

$$\theta(n+1) = \theta(n) + a(n)[\frac{\partial}{\partial\theta}\Phi(U_{n+1},\theta(n))],$$

which will track the o.d.e.

$$\dot{\theta}(t) = \frac{d}{d\theta}J(\theta).$$

This is the desired gradient ascent. This computation is the basic idea behind *infinitesimal perturbation analysis* (IPA) and its variants.

Another variation is the *likelihood ratio method* which assumes that the law μ_θ corresponding to F_θ is absolutely continuous with respect to a 'base probability measure' μ and the likelihood ratio (or *Radon–Nikodym derivative*) $\Lambda_\theta(\cdot) \stackrel{\text{def}}{=} \frac{d\mu_\theta}{d\mu}(\cdot)$ is continuously differentiable in θ. Then

$$J(\theta) = \int f d\mu_\theta = \int f\Lambda_\theta d\mu.$$

Suppose the interchange of expectation and differentiation

$$\frac{d}{d\theta}J(\theta) = \int f \frac{d}{d\theta}\Lambda_\theta d\mu$$

is justified. Then the stochastic approximation

$$\theta(n+1) = \theta(n) + a(n)[f(X_{n+1})\frac{d}{d\theta}\Lambda_\theta(X_{n+1})|_{\theta=\theta(n)}],$$

where $\{X_n\}$ are i.i.d. with law μ, will track the same o.d.e. as above.

It is also possible to conceive of a combination of the two schemes, see section 15.4 of Spall (2003). The methods get complicated if the $\{X_n\}$ are not independent, e.g., in case of a Markov chain. See Ho and Cao (1991), Glasserman (1991) and Fu and Hu (1997) for extensive accounts of IPA and its variants.

An alternative approach in case of a scalar parameter is to have two simulations $\{X_n\}$ and $\{X'_n\}$ corresponding to $\{\theta(n)\}$ and its 'small perturbation' $\{\theta(n)+\delta\}$ for some small $\delta > 0$ respectively. That is, the conditional law of X_n (resp. X'_n), given $X_m, X'_m, m < n$, and $\theta(m), m \leq n$, is $\mu_{\theta(n)}$ (resp. $\mu_{\theta(n)+\delta}$). The iteration scheme is

$$\theta(n+1) = \theta(n) + a(n)\left(\frac{f(X'_{n+1}) - f(X_{n+1})}{\delta}\right).$$

By the results of Chapter 5, this tracks the o.d.e.

$$\dot{\theta}(t) = \frac{J(\theta+\delta) - J(\theta)}{\delta},$$

the approximate gradient ascent. Bhatnagar and Borkar (1998) take this viewpoint with an additional stochastic approximation iteration on a faster timescale for explicit averaging of $\{f(X_n)\}$ and $\{f(X_n')\}$. This leads to more graceful behaviour. Bhatnagar and Borkar (1997) do the same with the two-timescale effect achieved, not through the choice of different stepsize schedules, but by performing the slow iteration along an appropriately chosen subsample of the time instants at which the fast iteration is updated. As already mentioned in Chapter 6, in actual experiments a judicious combination of the two was found to work better than either by itself. The advantage of these schemes is that they are no more difficult to implement and analyze for controlled Markov $\{X_n\}$ (resp. $\{X_n'\}$), wherein the conditional law of X_{n+1} (resp. X_{n+1}') given $X_m, X_m', \theta(m), m \leq n$, depends on $X_n, \theta(n)$ (resp. $X_n', \theta(n)$), than they are for $\{X_n\}$ (resp. $\{X_n'\}$) as above wherein the latter depends on $\theta(n)$ alone. The disadvantage on the other hand is that for d-dimensional parameters, one needs $(d+1)$ simulations, one corresponding to $\{\theta(n)\}$ and d corresponding to a δ-perturbation of each of its d components. Bhatnagar et al. (2003) work around this by combining the ideas above with SPSA.

Most applications of stochastic gradient methods tend to be for minimization of an appropriately defined measure of mean 'error' or 'discrepancy'. The mean square error and the relative (Kullback–Leibler) entropy are the two most popular instances. We have seen one example of the former in Chapter 1, where we discussed the problem of finding the optimal parameter β to minimize $E[\|Y_n - f_\beta(X_n)\|^2]$, where $(Y_i, X_i), i \geq 1$, are i.i.d. pairs of observations and $f_\beta(\cdot)$ is a parametrized family of functions. Most parameter tuning algorithms in the neural network literature are of this form, although some use the other, i.e., 'entropic' discrepancy measure. See Haykin (1999) for an overview. In particular, the celebrated *backpropagation* algorithm is a stochastic gradient scheme involving the gradient of a 'layered' composition of sums of nonlinear maps. The computation of the gradient is split into simple local computations using the chain rule of calculus.

An even older application is to adaptive signal processing (Haykin, 1991). More sophisticated variants appear in system identification where one tries to learn the dynamics of a stochastic dynamic system based on observed outputs (Ljung, 1999).

See also Fort and Pages (1995) and Kosmatopoulos and Christodoulou (1996) for an analysis of the Kohonen algorithm for *learning vector quantization* (LVQ) that seeks to find points $x_1, \ldots, x_k$ (say) so as to minimize

$$E[\min_{1 \leq i \leq k} \|x_k - Y\|^2],$$

given i.i.d. samples of Y. These help us to identify k 'clusters' in the observations, each identified with one of the x_i, with the understanding that every observation is associated with the nearest $x_i, 1 \leq i \leq k$.

Gradient and gradient-like schemes are guaranteed to converge to a local minimum at best. A scheme that ensures convergence to a *global* minimum *in probability* is *simulated annealing*, which we do not deal with here. It involves adding a slowly decreasing extraneous noise to 'push' the iterates out of local minima that are not global minima often enough that their eventual convergence to the set of global minima (in probability) is assured. See, e.g., Gelfand and Mitter (1991).

An alternative 'correlation-based' scheme that asymptotically behaves like a gradient scheme, but does not involve explicit gradient estimation, is the 'Alopex' algorithm analyzed in Sastry, Magesh and Unnikrishnan (2002).

10.3 Stochastic fixed point iterations

In this section we consider iterations of the type

$$x_{n+1} = x_n + a(n)[F(x_n) - x_n + M_{n+1}].\qquad(10.3.1)$$

That is, $h(x) = F(x) - x$ in our earlier notation. The idea is that $\{x_n\}$ should converge to a solution x^* of the equation $F(x^*) = x^*$, i.e., to a fixed point of $F(\cdot)$. We shall be interested in the following specific situation: Let $w_i > 0, 1 \leq i \leq d$, be prescribed 'weights' and define norms on $\mathcal{R}^d$ equivalent to the usual Euclidean norm: for $x = [x_1, \ldots, x_d] \in \mathcal{R}^d$ and $w = [w_1, \ldots, w_d]$ as above,

$$||x||_{w,p} \stackrel{\text{def}}{=} (\sum_{i=1}^{d} w_i |x_i|^p)^{\frac{1}{p}}, \ 1 \leq p < \infty,$$

and

$$||x||_{w,\infty} \stackrel{\text{def}}{=} \max_i w_i |x_i|.$$

We assume that

$$||F(x) - F(y)||_{w,p} \leq \alpha ||x - y||_{w,p} \ \forall \ x, y \in \mathcal{R}^d,\qquad(10.3.2)$$

for some w as above, $1 \leq p \leq \infty$, and $\alpha \in [0, 1]$. We shall say that F is a *contraction* w.r.t. the norm $|| \cdot ||_{w,p}$ if (10.3.2) holds with $\alpha \in [0, 1)$ and a *non-expansive map* w.r.t. this norm if it holds with $\alpha = 1$. As the names suggest, in the former case an application of the map contracts distances by a factor of at least $\alpha < 1$, in the latter case it does not increase or 'expand' them. By the contraction mapping theorem (see Appendix A), a contraction has a unique fixed point whereas a non-expansive map may have none (e.g., $F(x) = x + 1$), one (e.g., $F(x) = -x$), or many, possibly infinitely many (e.g., $F(x) = x$).

The limiting o.d.e. is

$$\dot{x}(t) = F(x(t)) - x(t). \tag{10.3.3}$$

We analyze this equation below for the case $\alpha \in [0,1)$, whence there is a unique x^* such that $x^* = F(x^*)$.

Let $F_i(\cdot)$ denote the ith component of $F(\cdot)$ for $1 \leq i \leq d$ and define $V(x) \overset{\text{def}}{=} \|x - x^*\|_{w,p}$, $x \in \mathcal{R}^d$.

Theorem 2. *Under (10.3.2), $V(x(t))$ is a strictly decreasing function of t for any non-constant trajectory of (10.3.3).*

Proof. Note that the only constant trajectory of (10.3.3) is $x(\cdot) \equiv x^*$. Let $1 < p < \infty$. Let $\mathrm{sgn}(x) = +1, -1$, or 0, depending on whether $x > 0, < 0$, or $= 0$. For $x(t) \neq x^*$, we obtain using the Hölder inequality,

$$
\begin{aligned}
&\frac{d}{dt}V(x(t)) \\
={}& \frac{p}{p}\Big(\sum_i w_i |x_i(t) - x_i^*|^p\Big)^{\frac{1-p}{p}}\Big(\sum_i w_i \mathrm{sgn}(x_i(t) - x_i^*) \\
&\qquad \times |x_i(t) - x_i^*|^{p-1}\dot{x}_i(t)\Big) \\
={}& \|x(t) - x^*\|_{w,p}^{1-p}\Big(\sum_i w_i \mathrm{sgn}(x_i(t) - x_i^*)|x_i(t) - x_i^*|^{p-1} \\
&\qquad \times (F_i(x(t)) - x_i(t))\Big) \\
={}& \|x(t) - x^*\|_{w,p}^{1-p}\Big(\sum_i w_i \mathrm{sgn}(x_i(t) - x_i^*)|x_i(t) - x_i^*|^{p-1} \\
&\qquad \times (F_i(x(t)) - F_i(x^*))\Big) \\
&- \|x(t) - x^*\|_{w,p}^{1-p}\Big(\sum_i w_i \mathrm{sgn}(x_i(t) - x_i^*)|x_i(t) - x_i^*|^{p-1} \\
&\qquad \times (x_i(t) - x_i^*)\Big) \\
\leq{}& \|x(t) - x^*\|_{w,p}^{1-p}\|x(t) - x^*\|_{w,p}^{p-1}\|F(x(t)) - F(x^*)\|_{w,p} \\
&- \|x(t) - x^*\|_{w,p}^{1-p}\|x(t) - x^*\|_{w,p}^{p-1}\|x(t) - x^*\|_{w,p} \\
={}& \|F(x(t)) - F(x^*)\|_{w,p} - \|x(t) - x^*\|_{w,p} \\
\leq{}& -(1 - \alpha)\|x(t) - x^*\|_{w,p} = -(1 - \alpha)V(x(t)),
\end{aligned}
$$

which is < 0 for $x(t) \neq x^*$. Here the first term on the right-hand side of the first inequality comes from the first term on its left-hand side by Hölder's inequality, and the second term on the right exactly equals the second term on the left. This proves the claim for $1 < p < \infty$. The inequality above can be written, for $t > s \geq 0$, as

$$\|x(t) - x^*\|_{w,p} \leq \|x(s) - x^*\|_{w,p} - (1 - \alpha)\int_s^t \|x(y) - x^*\|_{w,p}\,dy.$$

Letting $p \downarrow 1$ (resp. $p \uparrow \infty$), the claims for $p = 1$ (resp. $p = \infty$) follow by continuity of $p \to \|x\|_{w,p}$ on $[1, \infty]$. ∎

Corollary 3. x^* *is the unique globally asymptotically stable equilibrium of (10.3.3).*

Remark: This corollary extends to the non-autonomous case

$$\dot{x}(t) = F_t(x(t)) - x(t)$$

in a straightforward manner when the maps $F_t, t \geq 0$, satisfy

$$\|F_t(y) - F_t(z)\|_p \leq \alpha \|y - z\|_p \ \forall \ t \geq 0,$$

for $\alpha \in (0, 1)$, *and* have a common (unique) fixed point x^*.

The non-expansive case ($\alpha = 1$) for $p = \infty$ is sometimes useful in dynamic-programming-related applications. We state the result without proof here, referring the reader to Borkar and Soumyanath (1997) for a proof.

Theorem 4. *Let $F(\cdot)$ be $\| \cdot \|_{w,\infty}$-non-expansive. If $H \overset{def}{=} \{x : F(x) = x\} \neq \phi$, then $\|x(t) - x\|_{w,\infty}$ is nonincreasing for any $x \in H$ and $x(t) \to$ a single point in H (depending on $x(0)$).*

In particular, the corresponding stochastic approximation iterates converge to H a.s. by familiar arguments.

The most important application of this set-up is to the reinforcement learning algorithms for Markov decision processes. We shall illustrate a simple case here, that of the infinite-horizon discounted-cost Markov decision process. Thus we have a controlled Markov chain $\{X_n\}$ on a finite state space S, controlled by a control process $\{Z_n\}$ taking values in a finite 'action space' A. Its evolution is governed by

$$P(X_{n+1} = i | X_m, Z_m, m \leq n) = p(X_n, Z_n, i), \ n \geq 0, i \in S. \tag{10.3.4}$$

Here $p : S \times A \times S \to [0, 1]$ is the controlled transition probability function satisfying

$$\sum_j p(i, a, j) = 1 \ \forall i \in S, a \in A.$$

Thus $p(i, a, j)$ is the probability of moving from i to j when the action chosen in state i is a, regardless of the past. Let $k : S \times A \to \mathcal{R}$ be a prescribed 'running cost' function and $\beta \in (0, 1)$ a prescribed 'discount factor'. The classical discounted cost problem is to minimize

$$J(i, \{Z_n\}) \overset{def}{=} E[\sum_{m=0}^{\infty} \beta^m k(X_m, Z_m)|X_0 = i] \tag{10.3.5}$$

over admissible $\{Z_n\}$ (i.e., those that are consistent with (10.3.4)) for each i. The classical approach to this problem is through dynamic programming (see, e.g., Puterman, 1994). Define the 'value function'

$$V(i) \stackrel{\text{def}}{=} \inf_{\{Z_n\}} J(i, \{Z_n\}), \ i \in S.$$

It then satisfies the dynamic programming equation

$$V(i) = \min_a [k(i,a) + \beta \sum_j p(i,a,j)V(j)], \ i \in S. \qquad (10.3.6)$$

In words, this says that 'the minimum expected cost to go from state i is the minimum of the expected sum of the immediate cost at i *and* the minimum cost to go from the next state on'. Furthermore, for a $v : S \to A$, the control choice $Z_n = v(X_n) \ \forall n$ is optimal for all choices of the initial law for X_0 if $v(i)$ attains the minimum on the right-hand side of (10.3.6) for all i. Thus the problem is 'solved' if we know $V(\cdot)$. Note that (10.3.6) is of the form $V = F(V)$ for $F(\cdot) = [F_1(\cdot), \ldots, F_{|S|}(\cdot)]$ defined as follows: for $x = [x_1, \ldots, x_{|S|}] \in \mathcal{R}^{|S|}$,

$$F_i(x) \stackrel{\text{def}}{=} \min_a [k(i,a) + \beta \sum_j p(i,a,j)x_j], \ 1 \le i \le |S|.$$

It is easy to verify that

$$\|F(x) - F(y)\|_\infty \le \beta \|x - y\|_\infty, \ x, y \in \mathcal{R}^{|S|}.$$

Thus by the contraction mapping theorem (see Appendix A), equation (10.3.6) has a unique solution V and the 'fixed point iterations' $V_{n+1} = F(V_n), n \ge 0$, converge exponentially to V for any choice of V_0. These iterations constitute the well-known 'value iteration' algorithm, one of the classical computational schemes of Markov decision theory (see Puterman, 1994, for an extensive account).

The problem arises if the function $p(\cdot)$, i.e., the system model, is unknown. Thus the conditional averaging with respect to $p(\cdot)$ implicit in the evaluation of F cannot be performed. Suppose, however, that a simulation device is available which can generate a transition according to the desired conditional law $p(i, a, \cdot)$ (say). This situation occurs typically with large complex systems constructed by interconnecting several relatively simpler systems, so that while complete analytical modelling and analysis is unreasonably hard, a simulation based on 'local' rules at individual components is not (e.g., a communication network). One can then use the simulated transitions coupled with stochastic approximation to average their effects in order to mimic the value iteration. The 'simulation' can also be an actual run of the real system in the 'on-line' version of the algorithm.

This is the basis of the *Q-learning* algorithm of Watkins (1989), a cornerstone of reinforcement learning. Before we delve into its details, a word on the notion of reinforcement learning: In the classical learning paradigms for autonomous agents in artificial intelligence, one has at one end of the spectrum *supervised learning* in which *instructive feedback* such as the value of an error measure or its gradient is provided continuously to the agent as the basis on which to learn. This is the case, e.g., in the 'perceptron training algorithm' in neural networks – see Haykin (2000). (Think of a teacher.) At the other extreme are unsupervised learning schemes such as the learning vector quantization scheme of Kohonen (see, e.g., Kohonen, 2002) which 'self-organize' data without any external feedback. In between the two extremes lies the domain of *reinforcement learning* where the agent gets *evaluative feedback*, i.e., an observation related to the performance and therefore carrying useful information about it which, however, falls short of what would constitute exact instructive feedback. (Think of a critic.) In the context of Markov decision processes described above, the observed payoffs ($\approx$ negative of costs) constitute the reinforcement signal.

Q-learning derives its name from the fact that it works with the so-called Q-factors rather than with the value function. The Q-factors are nothing but the entities being minimized on the right-hand side of (10.3.6). Specifically, let

$$Q(i,a) \overset{\text{def}}{=} k(i,a) + \beta \sum_j p(i,a,j) V(j), \ i \in S, a \in A.$$

Thus in particular $V(i) = \min_a Q(i,a) \ \forall i$ and $Q(\cdot)$ satisfies

$$Q(i,a) = k(i,a) + \beta \sum_j p(i,a,j) \min_b Q(j,b), \ i \in S, a \in A. \qquad (10.3.7)$$

Like (10.3.6), this is also of the form $Q = G(Q)$ for a $G : \mathcal{R}^{|S| \times |A|} \to \mathcal{R}$ satisfying

$$\|G(Q) - G(Q')\|_\infty \le \beta \|Q - Q'\|_\infty.$$

In particular, (10.3.7) has a unique solution Q^* and the iteration $Q_{n+1} = G(Q_n)$ for $n \ge 0$, i.e.,

$$Q_{n+1}(i,a) = k(i,a) + \beta \sum_j p(i,a,j) \min_b Q_n(j,b), \ n \ge 0,$$

converges to Q^* at an exponential rate. What the passage from (10.3.6) to (10.3.7) has earned us is the fact that now the minimization is inside the conditional expectation and not outside, which makes a stochastic approximation version possible. The stochastic approximation version based on a simulation run (X_n, Z_n) governed by (10.3.4) is

$$Q_{n+1}(i, a)$$
$$= Q_n(i, a) + a(n)I\{X_n = i, Z_n = a\}$$
$$\times [k(i, a) + \beta \min_b Q_n(X_{n+1}, b) - Q_n(i, a)] \qquad (10.3.8)$$
$$\left(\; = (1 - a(n)I\{X_n = i, Z_n = a\})Q_n(i, a)\right.$$
$$+ a(n)I\{X_n = i, Z_n = a\}$$
$$\left.\times [k(i, a) + \beta \min_b Q_n(X_{n+1}, b)] \qquad\right)$$

for $n \geq 0$. Thus we replace the conditional expectation in the previous iteration by an actual evaluation at a random variable realized by the simulation device according to the desired transition probability in question, and then make an incremental move in that direction with a small weight $a(n)$, giving a large weight $(1 - a(n))$ to the previous guess. This makes it a stochastic approximation, albeit asynchronous, because only one component is being updated at a time. Note that the computation is still done by a single processor. Only one component is updated at a time purely because new information is available only for one component at a time, corresponding to the transition that just took place. Thus there is no reason to use $a(\nu(i, n))$ in place of $a(n)$ as in Chapter 7. The limiting o.d.e. then is

$$\dot{q}(t) = \Lambda(t)(G(q(t)) - q(t)),$$

where $\Lambda(t)$ for each t is a diagonal matrix with a probability vector along its diagonal. Assume that its diagonal elements remain uniformly bounded away from zero. Sufficient conditions that ensure this can be stated along the lines of section 7.4, viz., irreducibility of $\{X_n\}$ under all control policies of the type $Z_n = v(X_n), n \geq 0$, for some $v : S \to A$, *and* a requirement that at each visit to any state i, there is a minimum positive probability of choosing any control a in A. Then this is a special case of the situation discussed in example (ii) in section 7.4. As discussed there, the foregoing ensures convergence of the o.d.e. to Q^* and therefore the a.s. convergence of $\{Q_n\}$ to Q^*.

There are other reinforcement learning algorithms differing in philosophy (e.g., the actor-critic algorithm (Barto et al., 1983) which mimics the 'policy iteration' scheme of Markov decision theory), or differing in the cost criterion, e.g., the 'average cost' (Abounadi et al., 2001), or different because of an explicit approximation architecture incorporated to beat down the 'curse of dimensionality' (see, e.g., Tsitsiklis and Van Roy, 1997). They are all stochastic approximations.

Contractions and non-expansive maps are, however, not the only ones for which convergence to a unique fixed point may be proved. One other case, for example, is when $-F$ is *monotone*, i.e.,

$$\langle x - y, F(x) - F(y) \rangle < 0 \; \forall \; x \neq y.$$

In the affine case, i.e., $F(x) = Ax + b$ for a $d \times d$ matrix A and $b \in \mathcal{R}^d$, this would mean that the symmetric part of A, i.e., $\frac{1}{2}(A + A^T)$, would have to be negative definite. Suppose F has a fixed point x^*. Then

$$
\begin{aligned}
\frac{d}{dt} \|x(t) - x^*\|^2 &= 2\langle x(t) - x^*, F(x(t)) - x(t) \rangle \\
&= 2\langle x(t) - x^*, F(x(t)) - F(x^*) \rangle - 2\|x(t) - x^*\|^2 \\
&< 0
\end{aligned}
$$

for $x(t) \neq x^*$. Thus $\|x(t) - x^*\|^2$ serves as a Liapunov function, leading to $x(t) \to x^*$. In particular, if x' were another fixed point of F, $x(t) \equiv x'$ would satisfy (10.3.3), forcing $x' = x^*$. Hence x^* is the unique fixed point of F and a globally asymptotically stable equilibrium for the o.d.e.

10.4 Collective phenomena

The models we have considered so far are concerned with adaptation by a single agent. An exciting area of current research is the scenario when several interacting agents are each trying to adapt to an environment which in turn is affected by the other agents in the pool. A simple case is that of two agents in the 'nonlinear urn' scenario discussed in Chapter 1. Suppose we have two agents and an initially empty urn, with the first agent adding either zero or one red ball to the urn at each (discrete) time instant and the other doing likewise with black balls. Let x_n (resp. y_n) denote the *fraction of times up to time n that a red (resp. black) ball is added.* That is, if $\xi_n \overset{\text{def}}{=} I\{$ a red ball is added at time $n\}$ and $\zeta_n \overset{\text{def}}{=} I\{$ a black ball is added at time $n\}$ for $n \geq 1$, then

$$x_n \overset{\text{def}}{=} \frac{\sum_{m=1}^n \xi_m}{n}, \; y_n \overset{\text{def}}{=} \frac{\sum_{m=1}^n \zeta_m}{n}, \; n \geq 1.$$

Suppose the conditional probability that a red ball is added at time n given the past up to time n is $p(y_n)$ and the corresponding conditional probability that a black ball is added at time n is $q(x_n)$, for prescribed Lipschitz functions $p(\cdot), q(\cdot) : [0,1] \to [0,1]$. That is, the probability with which an agent adds a ball at any given time depends on the empirical frequency with which the other agent has been doing so until then. Arguing as in Chapter 1, we then have

$$
\begin{aligned}
x_{n+1} &= x_n + \frac{1}{n+1}[p(y_{n+1}) - x_n + M_{n+1}], \\
y_{n+1} &= y_n + \frac{1}{n+1}[q(x_{n+1}) - y_n + M'_{n+1}],
\end{aligned}
$$

for suitably defined martingale differences $\{M_{n+1}\}, \{M'_{n+1}\}$. This leads to the o.d.e.

$$\dot{x}(t) = p(y(t)) - x(t), \quad \dot{y}(t) = q(x(t)) - y(t), \quad t \geq 0.$$

Note that the 'driving vector field' on the right-hand side,

$$h(x,y) = [h_1(x,y), h_2(x,y)]^T \stackrel{\text{def}}{=} [p(y) - x, q(x) - y]^T,$$

satisfies

$$\mathrm{Div}(h) \stackrel{\text{def}}{=} \frac{\partial h_1(x,y)}{\partial x} + \frac{\partial h_2(x,y)}{\partial y} = -2.$$

From o.d.e. theory, one knows then that the maps $(x,y) \to (x(t), y(t))$ for $t > 0$ 'shrink' the volume in $\mathcal{R}^2$. In fact one can show that the flow becomes asymptotically one dimensional and therefore, as argued in Chapter 1, must converge (Benaim and Hirsch, 1997). That is, the fractions of red and black balls stabilize, leading to a fixed asymptotic probability of picking for either.

While such nonlinear urns have been studied as economic models, this analysis extends to more general problems, viz., two-person repeated bimatrix games (Kaniovski and Young, 1995). Here two agents, say agent 1 and agent 2, repeatedly play a game in which there are the same two strategy choices available to each of them at each time, say $\{a_1, b_1\}$ for agent 1 and $\{a_2, b_2\}$ for agent 2. Based on the strategy pair $(\xi_n^1, \xi_n^2) \in \{a_1, b_1\} \times \{a_2, b_2\}$ chosen at time n, agent i gets a payoff of $h_i(\xi_n^1, \xi_n^2)$ for $i = 1, 2$. Let

$$\nu_i(n) \stackrel{\text{def}}{=} \frac{\sum_{m=1}^n I\{\xi_m^i = a_i\}}{n}, \quad i = 1, 2; \; n \geq 0,$$

specify their respective 'empirical strategies'. That is, at time n agent i appears to agent $j \neq i$ as though she is choosing a_i with probability $\nu_i(n)$ and b_i with probability $1 - \nu_i(n)$. We assume that agent $j \neq i$ plays at time $n + 1$ her 'best response' to agent i's empirical strategy given by the probability vector $[\nu_i(n), 1 - \nu_i(n)]$. Suppose that this best response is given in turn by a Lipschitz function $f_j : [0,1] \to [0,1]$. That is, she plays $\xi_{n+1}^j = a_j$ with probability $f_j(\nu_i(n))$. Modulo the assumed regularity of the best response maps f_1, f_2, this is precisely the 'fictitious play' model of Brown (1951), perhaps the first learning model in game theory which has been extensively analyzed. A similar rule applies when i and j are interchanged. Then the above analysis leads to the limiting o.d.e.

$$\begin{aligned}
\dot{\nu}_1(t) &= F_1(\nu_1(t), \nu_2(t)) \stackrel{\text{def}}{=} f_1(\nu_2(t)) - \nu_1(t), \\
\dot{\nu}_2(t) &= F_2(\nu_1(t), \nu_2(t)) \stackrel{\text{def}}{=} f_2(\nu_1(t)) - \nu_2(t).
\end{aligned}$$

Again, $\mathrm{div}([F_1, F_2]) \stackrel{\text{def}}{=} \frac{\partial F_1}{\partial \nu_1} + \frac{\partial F_2}{\partial \nu_2} = -2$, leading to the same conclusion as before. That is, their strategies converge a.s. This limit, say $[\nu_1^*, \nu_2^*]$, forms a

Nash equilibrium: neither agent can improve her lot by moving away from the chosen strategy if the other one doesn't. This is immediate from the fact that ν_i^* is the best response to ν_j^* for $i \neq j$.

There are, however, some drastic oversimplifications in the foregoing. One important issue is that the 'best response' is often non-unique and thus one has to replace this o.d.e. by a suitable differential inclusion. Also, the situation in dimensions higher than two is no longer as easy. See Chapters 2 and 3 of Fudenberg and Levine (1998) for more on fictitious play.

Another model of interacting agents is the o.d.e.

$$\dot{x}(t) = h(x(t)) \tag{10.4.1}$$

with

$$\frac{\partial h_i}{\partial x_j} > 0, \; j \neq i. \tag{10.4.2}$$

These are called *cooperative* o.d.e.s, the idea being that increase in the jth component, corresponding to some desirable quantity for the jth agent, will lead to an increase in the ith component as well for $i \neq j$. (The strict inequality in (10.4.2) can be weakened to '$\geq$' as long as the Jacobian matrix of h is irreducible, i.e., for any partition $I \cup J$ of the row/column indices, there is some $i \in I, j \in J$ such that the (i,j)th element is nonzero. See section 4.1 of Smith, 1995, for details.) Suppose the trajectories remain bounded. Then a theorem of Hirsch states that for all initial conditions belonging to an open dense set, $x(\cdot)$ converges to the set of equilibria (see Hirsch, 1985; also Smith, 1995).

As an application, we consider the problem of dynamic pricing in a system of parallel queues from Borkar and Manjunath (2004). There are K parallel queues and an entry charge $p_i(n)$ is charged for the ith queue, reflecting (say) its quality of service. Let $y_i(n)$ denote the queue length in the ith queue at time n. There is an 'ideal profile' $y^* = [y_1^*, \ldots, y_K^*]$ of queue lengths which we want to stay close to, and the objective is to manage this by modulating the respective prices dynamically. Let Γ_i denote the projection to $[\epsilon_i, B_i]$ for $0 \leq i \leq K$, where $\epsilon_i > 0$ is a small number and B_i is a convenient a priori upper bound. Let $a > 0$ be a small constant stepsize. The scheme is, for $1 \leq i \leq K$,

$$p_i(n+1) = \Gamma_i \left(p_i(n) + a p_i(n)[y_i(n) - y_i^*] \right), \; n \geq 0.$$

The idea is to increase the price if the current queue length is above the ideal (so as to discourage new entrants) and decrease it if the opposite is true (to encourage more entrants). The scalar ϵ_i is the minimum price which also ensures that the iteration does not get stuck at zero. We assume that if the price vector is frozen at some $p = [p_1, \ldots, p_K]$, the process of queue lengths is ergodic. Ignoring the boundary of the box $\mathcal{B} \stackrel{\text{def}}{=} \Pi_i[\epsilon_i, B_i]$, the limiting o.d.e. is

$$\dot{p}_i(t) = p_i(t)[f_i(p(t)) - y_i^*], \ 1 \leq i \leq K,$$

where $p(t) = [p_1(t), \ldots, p_K(t)]$ and $f_i(p)$ is the stationary average of the queue length in the ith queue when the price vector is frozen at p. It is reasonable to assume that if the ϵ_i are sufficiently low and the B_i are sufficiently high, then $p(t)$ is eventually pushed inwards from the boundary of $\mathcal{B}$, so that we may ignore the boundary effects and the above is indeed the valid o.d.e. limit for the price adjustment mechanism. It is also reasonable to assume that

$$\frac{\partial f_i}{\partial p_j} > 0,$$

as an increase in the price of one queue keeping all else constant will force its potential customers to other queues. (As mentioned above, this condition can be relaxed.) Thus this is a cooperative o.d.e. and the foregoing applies. One can say more: Letting $f(\cdot) = [f_1(\cdot), \ldots, f_K(\cdot)]^{\mathrm{T}}$, it follows by Sard's theorem (see, e.g., Ortega and Rheinboldt, 1970, p. 130) that for almost all choices of y^*, the Jacobian matrix of f is nonsingular on the inverse image of y^*. Hence by the inverse function theorem (see, e.g., *ibid.*, p. 125), this set, which is also the set of equilibria for the above o.d.e. in $\mathcal{B}$, is discrete. Thus the o.d.e. converges to a point for almost all initial conditions. One can argue as in Chapter 9 to conclude that the stationary distribution of $\{p(n)\}$ concentrates near the equilibria of the above o.d.e. Note that all equilibria in $\mathcal{B}$ are equivalent as far as our aim of keeping the vector of queue lengths near y^* is concerned. Thus we conclude that the dynamically adjusted prices asymptotically achieve the desired queue length profile, giving it the right 'pushes' if it deviates.

Yet another popular dynamic model of interaction is the celebrated *replicator dynamics*, studied extensively by mathematical biologists – see, e.g., Hofbauer and Siegmund (1998). This dynamics resides in the d-dimensional probability simplex $S \stackrel{\text{def}}{=} \{x = [x_1, \ldots, x_d] : x_i \geq 0 \ \forall i, \ \sum_i x_i = 1\}$, and is given by

$$\dot{x}_i(t) = x_i(t)[D_i(x(t)) - \sum_j x_j(t)D_j(x(t))], \ 1 \leq i \leq d, \qquad (10.4.3)$$

where $x(t) = [x_1(t), \ldots, x_d(t)]$ and $D(\cdot) = [D_1(\cdot), \ldots, D_d(\cdot)]$ is a Lipschitz map. The interpretation is that for each i, $x_i(t)$ is the fraction of the population at time t that belongs to species i. $D_i(x)$ is the payoff received by the ith species when the population profile is x (i.e., when the fraction of the population occupied by the jth species is x_j for all j). Equation (10.4.3) then says in particular that the fraction of a given species in the population increases if the payoff it is receiving is higher than the population average, and decreases if it is lower. If $\sum_i x_i(t) = 1$, then summed over i, the right-hand side of (10.4.3) vanishes, implying that $\sum_i x_i(t) = 1 \ \forall t$ if it is so for $t = 0$. Thus the simplex S is invariant under this o.d.e. The faces of S correspond to one or more components being zero, i.e., the population of that particular species is 'extinct'. These are also individually invariant, because the corresponding components of the right-hand side of (10.4.3) vanish.

This equation has been studied extensively (Hofbauer and Siegmund, 1998). We shall consider it under the additional condition that $-D(\cdot)$ be *monotone*, i.e.,

$$\langle x - y, D(x) - D(y) \rangle < 0 \ \forall \ x \neq y \in S.$$

In the affine case, i.e., $D(x) = Ax + b$ for a $d \times d$ matrix A and $b \in \mathcal{R}^d$, this would mean that the symmetric part of A, i.e., $\frac{1}{2}(A + A^{\mathrm{T}})$, would have to be negative definite.

Lemma 5. *There exists a unique $x^* \in S$ such that x^* maximizes $x \to x^{\mathrm{T}} D(x^*)$.*

Proof. The set-valued map that maps $x \in S$ to the set of maximizers of the function $y \in S \to y^{\mathrm{T}} D(x)$ is nonempty compact convex and upper semicontinuous, as can be easily verified. Thus by the Kakutani fixed point theorem (see, e.g., Border, 1985), it follows that there exists an x^* such that x^* maximizes the function $y \to y^{\mathrm{T}} D(x^*)$ over S. Suppose $x' \neq x^*$ is another point in S such that x' maximizes the function $y \to y^{\mathrm{T}} D(x')$ on S. Then

$$\langle x^* - x', D(x^*) - D(x') \rangle$$
$$= (x^*)^{\mathrm{T}} D(x^*) - (x^*)^{\mathrm{T}} D(x') - (x')^{\mathrm{T}} D(x^*) + (x')^{\mathrm{T}} D(x')$$
$$\geq 0.$$

This contradicts the 'monotonicity' condition above unless $x^* = x'$. ∎

Theorem 6. *For $x(0)$ in the interior of S, $x(t) \to x^*$.*

Proof. Define

$$V(x) \stackrel{\mathrm{def}}{=} \sum_i x_i^* \ell n \left(\frac{x_i^*}{x_i} \right), \quad x = [x_1, \cdots, x_d] \in S.$$

Then an application of Jensen's inequality shows that $V(x) \geq 0$ and is 0 if and only if $x = x^*$. Also,

$$\frac{d}{dt} V(x(t)) = -\sum_i x_i^* \left(\frac{\dot{x}_i(t)}{x_i(t)} \right)$$
$$= (x(t) - x^*)^{\mathrm{T}} D(x(t))$$
$$\leq (x(t) - x^*)^{\mathrm{T}} D(x(t)) - (x(t) - x^*)^{\mathrm{T}} D(x^*)$$
$$= (x(t) - x^*)^{\mathrm{T}} (D(x(t)) - D(x^*))$$
$$< 0,$$

for $x(t) \neq x^*$. Here the second equality follows from the o.d.e. (10.4.3), the first inequality follows from our choice of x^*, and the last inequality follows from the monotonicity assumption on $-D$. The claim follows easily from this. ∎

Thus the stochastic approximation counterpart will converge a.s. to x^*. Note, however, that one requires a projection to keep the iterates in S, and also 'sufficiently rich' noise, possibly achieved by adding some extraneous noise, in order to escape getting trapped in an undesirable face of S.

Another 'convergent' scenario is when $D_j(x) = \frac{\partial F}{\partial x_j}$ $\forall j$ for some continuously differentiable function $F(\cdot)$. In that case,

$$\frac{d}{dt}F(x(t)) = \sum_i x_i(t) \left(\frac{\partial F}{\partial x_i}\right)^2 - \left(\sum_i x_i \frac{\partial F}{\partial x_i}\right)^2 \geq 0,$$

with strict inequality when $\nabla F \neq 0$, so that $-F(\cdot)$ serves as a Liapunov function. See Sandholm (1998) for an interesting application of this to transportation science. A special case arises when $D_j(x) = \sum_k R(i,j)x_j$ $\forall j$, where R is a positive definite matrix. A time-scaled version of this occurs in the analysis of the asymptotic behaviour of vertex-reinforced random walks (Benaim, 1997). Pemantle (2006) surveys this and related dynamics.

The monotonicity or gradient conditions above, however, are not natural in many applications. Without monotonicity, (10.4.3) can show highly complex behaviour (Hofbauer and Siegmund, 1998). In a specific application, Borkar and Kumar (2003) get a partial characterization of the asymptotic behaviour.

The reinforcement learning paradigm introduced in section 10.3 also has a multiagent counterpart in which several agents control the transition probabilities of the Markov chain, each with its own objective in mind. The special case of this when a fixed static game is played repeatedly is called a 'repeated game'. We have already seen one instance of this situation in our discussion of fictitious play. This particular scenario has been extensively studied. Nevertheless, the results are limited except for special cases. The case of a two-person zero-sum game, wherein one agent tries to maximize a performance measure while the other agent tries to minimize it, is one situation where the dynamic programming ideas can be extended in a straightforward way. For example, suppose this performance measure is of the form (10.3.5) with Z_n standing for a *pair* of controls (Z_n^1, Z_n^2) chosen respectively by the two agents independently of each other. Then (10.3.6) simply extends to the *Shapley equation*

$$V(i) = \min_a \max_b [k(i,(a,b)) + \beta \sum_j p(i,(a,b),j)V(j)]$$

$$= \max_b \min_a [k(i,(a,b)) + \beta \sum_j p(i,(a,b),j)V(j)], \quad i \in S.$$

The corresponding Q-learning scheme may then be written accordingly. See

Hu and Wellman (1998), Jafari et al. (2001), Littman (2001), Sastry et al. (1994), Singh et al. (2000), for some important contributions to this area, which also highlight the difficulties. Leslie and Collins (2002) provide an interesting example of multiple timescales in this context. Despite all this and much else, the area of multiagent learning remains wide open with several unsolved issues. See Fudenberg and Levine (1998), Vega Redondo (1996), Young (1998), Young (2004) for a flavour of some of the economics-motivated work in this direction and Sargent (1993) for some more applications of stochastic approximation in economics, albeit with a different flavour.

10.5 Miscellaneous applications

This section discusses a couple of instances which don't fall into the above categorization, just to emphasize that the rich possibilities extend well beyond the paradigms described above.

(i) *A network example:* Consider the o.d.e.

$$\dot{x}_i(t) = a_i(x(t))\Big[b_i(x_i(t)) - \sum_j c_{ij} f_j(\sum_k c_{jk} g_k(x_k(t)))\Big].$$

Assume:

- $a_i(x), b_i(y), f_i(y), g_i(y), g_i'(y)$ are strictly positive and are Lipschitz, where $g_i' = \frac{dg_i}{dy}$.
- $c_{ij} = c_{ji}$.

Then $V(x) = \sum_i \Big(\int_0^{\sum_j c_{ij} g_j(x_j)} f_i(y)dy - \int_0^{x_i} b_i(y)g_i'(y)dy \Big)$ serves as a Liapunov function, because

$$\begin{aligned}
dV&(x(t))/dt \\
&= \sum_i[\sum_j f_j(\sum_k c_{jk} g_k(x_k(t)))c_{ji}g_i'(x_i(t)) \\
&\quad - b_i(x_i(t))g_i'(x_i(t))]\dot{x}_i(t) \\
&= -\sum_i a_i(x(t))g_i'(x_i(t))[b_i(x_i(t)) \\
&\quad - \sum_j c_{ij} f_j(\sum_k c_{jk} g_k(x_k(t)))]^2 \\
&\leq 0.
\end{aligned}$$

One can have 'neighbourhood structure' by having $N(i) \overset{\text{def}}{=}$ the set of neighbours of i, with the requirement that

$$c_{ij} = c_{ji} > 0 \Longleftrightarrow i \in N(j) \Longleftrightarrow j \in N(i).$$

One can allow $i \in N(i)$. One 'network' interpretation is:

- $j \in N(i)$ if j's transmission can be 'heard' by i.
- $x_j(t) =$ the traffic originating from j at time t.
- $\{c_{ij}\}$ capture the distance effects.
- $\sum_k c_{jk} g_k(x_k(t)) =$ the net traffic 'heard' by j at time t.
- Each node reports the volume of the net traffic it has heard to its neighbours and updates its own traffic so that the higher the net traffic it hears, the more it decreases its own flow correspondingly.

An equilibrium of this o.d.e. will be characterized by

$$b_i(x_i) = \sum_j c_{ij} f_j \left(\sum_k c_{jk} g_k(x_k) \right) \forall i.$$

While we have used a network interpretation here, one can also recover the Kelly–Maulloo–Tan (1998) model of the *tatonnement* process in network pricing context and the Cohen–Grossberg model (see Chapter 14 of Haykin, 2000) which covers several neural network models including the celebrated Hopfield model. An important precursor to the Cohen–Grossberg model is Grossberg (1978).

(ii) *Principal component analysis:* This is a problem from statistics, wherein one has n-dimensional data for large $n > 1$ and the idea is to find $m << n$ directions along which the data can be said to be concentrated. Standard theoretical considerations then suggest that this should be the eigenspace of the empirical covariance matrix corresponding to its m largest eigenvalues. The neural network methodology for finding this subspace is based on adaptively learning the weight matrix of an appropriately designed network using stochastic approximation. There are several variations on this theme, see, e.g., section 8.3 of Hertz, Krogh and Palmer (1991). One celebrated instance due to Oja (1982) leads to the limiting matrix o.d.e. in $\mathcal{R}^{n \times m}$ given by

$$\dot{W}(t) = QW(t) - W(t)(W(t)^{\mathrm{T}} QW(t)),$$

where $Q \in \mathcal{R}^{n \times n}$ is a positive definite matrix. (Assuming zero mean data, this will in fact be its covariance matrix obtained from empirical data by the averaging property of stochastic approximation.) Leizarowitz (1997) analyzes this equation by control theoretic methods and establishes that $W(t)$ does indeed converge to a matrix whose column vectors span the eigenspace corresponding to the m largest eigenvalues of Q (see also Yan, Helmke and Moore, 1994). See Oja and Karhunen (1985) for a precursor and Yoshizawa, Helmke and Moore (2001), Helmke and Moore (1993) for yet another related dynamics with several important applications.

The account above is far from exhaustive. (Just to give one example, we have not covered the very interesting dynamics for consensus / coordination being studied in the robotics and sensor networks literature – see, e.g., Cucker and Smale (2007) and the references therein.) The message, if any, of the foregoing is that any convergent o.d.e. is a potential paradigm for a stochastic-approximation-based algorithm.

11
Appendices

11.1 Appendix A: Topics in analysis

11.1.1 Continuous functions

We briefly recall here the two key theorems about continuous functions used in the book. Recall that a subset of a topological space is relatively compact (resp. relatively sequentially compact) if its closure is compact (resp. sequentially compact). Also, compactness and sequential compactness are equivalent notions for metric spaces. The first theorem concerns the relative compactness in the space $C([0,T];\mathcal{R}^d)$ of continuous functions $[0,T] \to \mathcal{R}^d$ for a prescribed $T > 0$. $C([0,T];\mathcal{R}^d)$ is a Banach space under the 'sup-norm' $\|f\| \stackrel{\text{def}}{=} \sup_{t\in[0,T]} \|f(t)\|$. That is,

(i) it is a vector space over the reals,

(ii) $\| \cdot \| : C([0,T];\mathcal{R}^d) \to [0,\infty)$ satisfies

    (a) $\|f\| \geq 0$, with equality if and only if $f \equiv 0$,

    (b) $\|\alpha f\| = |\alpha| \|f\|$ for $\alpha \in \mathcal{R}$,

    (c) $\|f + g\| \leq \|f\| + \|g\|$,

  and

(iii) $\| \cdot \|$ is *complete*, i.e., $\{f_k\} \subset C([0,T];\mathcal{R}^d)$, $\|f_m - f_n\| \to 0$ as $m, n \to \infty$, imply that there exists an $f \in C([0,T];\mathcal{R}^d)$ such that $\|f_n - f\| \to 0$ (the uniqueness of this f is obvious).

A set $B \subset C([0,T];\mathcal{R}^d)$ is said to be *equicontinuous* at $t \in [0,T]$ if for any $\epsilon > 0$, one can find a $\delta > 0$ such that $|t - s| < \delta$, $s \in [0,T]$ implies $\sup_{f \in B} \|f(s) - f(t)\| < \epsilon$. It is simply *equicontinuous* if it equicontinuous at all $t \in [0,T]$. It is said to be *pointwise bounded* if for any $t \in [0,T]$, $\sup_{f\in B} \|f(t)\| < \infty$. It can be verified that if the set is equicontinuous, it will be pointwise bounded at all points in $[0,T]$ if it is so at one point in $[0,T]$. The result we are interested in is the *Arzela–Ascoli theorem* which characterizes relative compactness in $C([0,T];\mathcal{R}^d)$:

Theorem 1. *$B \subset C([0,T]; \mathcal{R}^d)$ is relatively compact if and only if it is equicontinuous and pointwise bounded.*

See Appendix A of Rudin (1991) for a proof and further developments.

The space $C([0,\infty); \mathcal{R}^d)$ of continuous functions $[0,\infty) \to \mathcal{R}^d$ is given the coarsest topology such that the map that takes $f \in C([0,\infty); \mathcal{R}^d)$ to its restriction to $[0,T]$, viewed as an element of the space $C([0,T]; \mathcal{R}^d)$, is continuous for all $T > 0$. In other words, $f_n \to f$ in this space if and only if $f_n|_{[0,T]} \to f|_{[0,T]}$ in $C([0,T]; \mathcal{R}^d)$ for all $T > 0$. This is not a Banach space, but a *Frechet* space, i.e., it has a complete translation-invariant metric and the corresponding open balls are convex. This metric can be, e.g.,

$$\rho(f,g) \stackrel{\text{def}}{=} \sum_{T=1}^{\infty} 2^{-T} \|f - g\|_T \wedge 1,$$

where we denote by $\| \cdot \|_T$ the sup-norm on $C([0,T]; \mathcal{R}^d)$ to make explicit its dependence on T. By our choice of the topology on $C([0,\infty); \mathcal{R}^d)$, Theorem 1 holds for this space as well.

The next result concerns *contractions*, i.e., maps $f : S \to S$ on a metric space S endowed with a metric ρ, that satisfy

$$\rho(f(x), f(y)) \leq \alpha \rho(x,y)$$

for some $\alpha \in [0,1)$. We say that x^* is a fixed point of f if $x^* = f(x^*)$. Assume that ρ is complete, i.e., $\lim_{m,n \to \infty} \rho(x_n, x_m) = 0$ implies $x_n \to x^*$ for some $x^* \in S$. The next theorem is called the *contraction mapping theorem*.

Theorem 2. *There exists a unique fixed point x^* of f and for any $x_0 \in S$, the iteration*

$$x_{n+1} = f(x_n), \ n \geq 0,$$

satisfies

$$\rho(x_n, x^*) \leq \alpha^n \rho(x_0, x^*), \ n \geq 0,$$

i.e., $\{x_n\}$ converges to x^* at an exponential rate.

This is an example of a *fixed point theorem*. Another example is *Brouwer's fixed point theorem*, which says that every continuous map $f : C \to C$ for a compact convex $C \subset \mathcal{R}^d$ has a fixed point.

11.1.2 Square-integrable functions

Consider now the space $L_2([0,T]; \mathcal{R}^d)$ of measurable functions $f : [0,T] \to \mathcal{R}^d$ satisfying

$$\int_0^T \|f(t)\|^2 dt < \infty.$$

Letting $\langle \cdot, \cdot \rangle$ denote the inner product in $\mathcal{R}^d$, we can define an inner product $\langle \cdot, \cdot \rangle_T$ on $L_2([0, T]; \mathcal{R}^d)$ by

$$\langle f, g \rangle_T \overset{\text{def}}{=} \int_0^T \langle f(t), g(t) \rangle dt, \ f, g \in L_2([0, T]; \mathcal{R}^d).$$

This is easily seen to be a valid inner product, i.e., a symmetric continuous map $L_2([0, T]; \mathcal{R}^d)^2 \to \mathcal{R}$ that is separately linear in each argument and satisfies: $\langle f, f \rangle_T \geq 0$, with equality if and only if $f \equiv 0$ a.e. It thus defines a norm

$$\|f\| \overset{\text{def}}{=} \sqrt{\langle f, f \rangle_T} = \left(\int_0^T \|f(t)\|^2 dt \right)^{\frac{1}{2}},$$

which turns out to be complete. $L_2([0, T]; \mathcal{R}^d)$ is then a *Hilbert* space with the above inner product and norm.

The open balls w.r.t. this norm define what is called the *strong* topology on $L_2([0, T]; \mathcal{R}^d)$. One can also define the *weak* topology as the coarsest topology w.r.t. which the functions $f \to \langle f, g \rangle_T$ are continuous for all $g \in L_2([0, T]; \mathcal{R}^d)$. The corresponding convergence concept is: $f_n \to f$ weakly in $L_2([0, T]; \mathcal{R}^d)$ if and only if $\langle f_n, g \rangle_T \to \langle f, g \rangle_T$ for all $g \in L_2([0, T]; \mathcal{R}^d)$. The results we need are the following:

Theorem 3. *A $\| \cdot \|$-bounded set $B \subset L_2([0, T]; \mathcal{R}^d)$ is relatively compact and relatively sequentially compact in the weak topology.*

Theorem 4. *If $f_n \to f$ weakly in $L_2([0, T]; \mathcal{R}^d)$, then there exists a subsequence $\{f_{n(k)}\}$ such that*

$$\left\| \frac{1}{m} \sum_{k=1}^m f_{n(k)} - f \right\| \to 0.$$

Theorem 3 is a special instance of the Banach–Alaoglu theorem. See Rudin (1991) for this and related developments. Likewise, Theorem 4 is an instance of the Banach–Saks theorem, see Balakrishnan (1976, p. 29) for a proof.

Let $\mathcal{X}$ denote the space of measurable maps $f : [0, \infty) \to \mathcal{R}^d$ with the property

$$\int_0^T \|f(t)\|^2 dt < \infty \ \forall T > 0.$$

Topologize $\mathcal{X}$ with the coarsest topology that renders continuous the maps

$$f \to \int_0^T \langle f(t)g(t) \rangle dt$$

for all $g \in L_2([0, T]; \mathcal{R}^d)$, for all $T > 0$. Then by our very choice of topology, Theorems 3 and 4 apply to $\mathcal{X}$ after the following modification: A set $B \subset \mathcal{X}$ with the property that $\{f|_{[0,T]} : f \in B\}$ is $\| \cdot \|_T$-bounded in $L_2([0, T]; \mathcal{R}^d)$ for all $T > 0$ will be relatively compact and relatively sequentially compact in $\mathcal{X}$. Furthermore, if $f_n \to f$ in $\mathcal{X}$, then for any $T > 0$, there exists a subsequence $\{f_{n(k)}\}$ such that

$$\|\frac{1}{m}\sum_{k=1}^{m} f_{n(k)}|_{[0,T]} - f|_{[0,T]}\|_T \to 0.$$

11.1.3 Lebesgue's theorem

Let $f : \mathcal{R} \to \mathcal{R}$ be a measurable and locally integrable function and for $t > s$ in $\mathcal{R}$, let $g(t) = \int_s^t f(y)dy$. Then Lebesgue's theorem states that for a.e. $t \geq s$, $\frac{d}{dt}g(t)$ exists and equals $f(t)$.

11.2 Appendix B: Ordinary differential equations

11.2.1 Basic theory

This chapter briefly summarizes some key facts about ordinary differential equations of relevance to us. The reader may refer to standard texts such as Hirsch, Smale and Devaney (2003) for further details. Consider the differential equation in $\mathcal{R}^d$ given by

$$\dot{x}(t) = h(x(t)), \ x(0) = \bar{x}. \tag{11.2.1}$$

This is an *autonomous* o.d.e. because the the driving vector field h does not have an explicit time-dependence. It would be *non-autonomous* if we replace $h(x(t))$ on the right by $h(x(t), t)$. We shall say that (11.2.1) is well-posed if for any choice of the initial condition $\bar{x} \in \mathcal{R}^d$, it has a unique solution $x(\cdot)$ defined for all $t \geq 0$ and the map $\bar{x} \to$ the corresponding $x(\cdot) \in C([0, \infty); \mathcal{R}^d)$ is continuous. One sufficient condition for this is the *Lipschitz condition* on $h(\cdot)$: there exists $L > 0$ such that

$$\|h(x) - h(y)\| \leq L\|x - y\|, \ \forall x, y \in \mathcal{R}^d.$$

Theorem 5. *For h satisfying the Lipschitz condition, (11.2.1) is well-posed.*

We shall sketch a proof of this to illustrate the application of the Gronwall inequality stated below:

Lemma 6. *(Gronwall inequality)* *For continuous $u(\cdot), v(\cdot) \geq 0$ and scalars $C, K, T \geq 0$,*

$$u(t) \leq C + K\int_0^t u(s)v(s)ds \ \forall t \in [0, T], \tag{11.2.2}$$

implies

$$u(t) \leq Ce^{K\int_0^T v(s)ds}, \ t \in [0, T].$$

Proof. Let $s(t) \overset{\text{def}}{=} \int_0^t u(s)v(s)ds$, $t \in [0,T]$. Multiplying (11.2.2) on both sides by $v(t)$, it translates into

$$\dot{s}(t) \leq Cv(t) + Ks(t)v(t).$$

This leads to

$$e^{-K\int_0^t v(s)ds}(\dot{s}(t) - Kv(t)s(t)) = \frac{d}{dt}\left(e^{-K\int_0^t v(s)ds}s(t)\right)$$
$$\leq Ce^{-K\int_0^t v(s)ds}v(t).$$

Integrating from 0 to t and using the fact that $s(0) = 0$, we have

$$e^{-K\int_0^t v(s)ds}s(t) \leq \frac{C}{K}(1 - e^{-K\int_0^t v(s)ds}).$$

Thus

$$s(t) \leq \frac{C}{K}(e^{K\int_0^t v(s)ds} - 1).$$

Hence

$$u(t) \leq C + Ks(t)$$
$$\leq C + K\left(\frac{C}{K}(e^{K\int_0^t v(s)ds} - 1)\right)$$
$$= Ce^{K\int_0^t v(s)ds}.$$

The claim follows for $t \in [0,T]$. ∎

The most commonly used situation is $v(\cdot) \equiv 1$, when this inequality reduces to

$$u(t) \leq Ce^{Kt}.$$

We return to the proof of Theorem 5.

Proof. Define the map $F : y(\cdot) \in C([0,T];\mathcal{R}^d) \to z(\cdot) \in C([0,T];\mathcal{R}^d)$ by

$$z(t) = \bar{x} + \int_0^t h(y(s))ds, \quad t \in [0,T].$$

Clearly, $x(\cdot)$ is a solution of (11.2.1) on $[0,T]$ if and only if it is a fixed point of F. Let $z_i(\cdot) = F(y_i(\cdot))$ for $i = 1,2$. Denoting by $\|\cdot\|_T$ the sup-norm on $C([0,T];\mathcal{R}^d)$, we have

$$\|z_1(\cdot) - z_2(\cdot)\|_T \leq \int_0^T \|h(y_1(s)) - h(y_2(s))\|ds$$
$$\leq L\int_0^T \|y_1(s) - y_2(s)\|ds$$
$$\leq LT\|y_1(\cdot) - y_2(\cdot)\|_T.$$

Taking $T < 1/L$, it follows that F is a contraction and thus has a unique fixed point by the contraction mapping theorem of Appendix A. Existence and uniqueness of a solution to (11.2.1) on $[0, T]$ follows. The argument may be repeated for $[T, 2T], [2T, 3T]$ and so forth in order to extend the claim to a general $T > 0$. Next, let $x_i(\cdot), i = 1, 2$, be solutions to (11.2.1) corresponding to $\bar{x} = \tilde{x}_1, \tilde{x}_2$, respectively. Then

$$\|x_1(t) - x_2(t)\| \leq \|\tilde{x}_1 - \tilde{x}_2\| + L \int_0^t \|x_1(s) - x_2(s)\| ds$$

for $t \in [0, T]$. By Lemma 6,

$$\|x_1(\cdot) - x_2(\cdot)\|_T \leq e^{LT} \|\tilde{x}_1 - \tilde{x}_2\|_T,$$

implying that the map $\bar{x} \in \mathcal{R}^d \to x(\cdot)|_{[0,T]} \in C([0, T]; \mathcal{R}^d)$ is Lipschitz, in particular, continuous. Since $T > 0$ was arbitrary, it follows that the map $\bar{x} \in \mathcal{R}^d \to C([0, \infty); \mathcal{R}^d)$ is continuous. ∎

Since the continuous image of a compact set is compact, we have:

Corollary 7. *The solution set of (11.2.1) as $\bar{x}$ varies over a compact subset of $\mathcal{R}^d$ is compact in $C([0, \infty); \mathcal{R}^d)$.*

A similar argument works if we consider (11.2.1) with $t \leq 0$: one simply has to work with intervals $[-T, 0]$ in place of $[0, T]$. Thus for each $t \in \mathcal{R}$, there is a continuous map $\Psi_t : \mathcal{R}^d \to \mathcal{R}^d$ that takes $\bar{x}$ to $x(t)$ via (11.2.1). It follows from the uniqueness claim above that Ψ_t, Ψ_{-t} are inverses of each other and thus each Ψ_t is a homeomorphism, i.e., a continuous bijection with a continuous inverse. The family $\Psi_t, t \in \mathcal{R}$, defines a *flow* of homeomorphisms $\mathcal{R}^d \to \mathcal{R}^d$, i.e., it satisfies:

  (i) $\Psi_0 = $ the identity map,
  (ii) $\Psi_s \circ \Psi_t = \Psi_t \circ \Psi_s = \Psi_{s+t}$,

where 'o' stands for composition of functions.

More generally, we may assume h to be only locally Lipschitz, i.e.,

$$\|h(x) - h(y)\| \leq L_R \|x - y\|, \ \forall x, y \in B_R \stackrel{\text{def}}{=} \{z \in \mathcal{R}^d : \|z\| \leq R\},$$

for some $L_R > 0$ that may tend to ∞ as $R \to \infty$. Then the claims of Theorem 5 can be shown to hold *locally* in space and time, i.e., in a neighbourhood of $\bar{x}$ and for $t \in [0, T]$ for $T > 0$ sufficiently small. Suppose in addition one can show separately that the trajectory is well-defined for all $t \geq 0$, i.e., there is no 'finite time blow-up' (meaning that $\lim_{t \uparrow \hat{t}} \|x(t)\| = \infty$ for some $\hat{t} < \infty$) for any initial condition. Then the full statement of Theorem 5 may be recovered. One way of ensuring no finite time blow-up is by demonstrating a convenient 'Liapunov function' – see section 11.3.

We close this section with a discrete counterpart of the Gronwall inequality. While it is not used much in the o.d.e. context, it is extremely useful otherwise and has been used extensively in this book itself.

Lemma 8. *(Discrete Gronwall inequality)* Let $\{x_n, n \geq 0\}$ (resp. $\{a_n, n \geq 0\}$) be nonnegative (resp. positive) sequences and $C, L \geq 0$ scalars such that for all n,

$$x_{n+1} \leq C + L(\sum_{m=0}^{n} a_m x_m). \tag{11.2.3}$$

Then for $T_n = \sum_{m=0}^{n} a_m$,

$$x_{n+1} \leq Ce^{LT_n}.$$

Proof. Let $s_n = \sum_{m=0}^{n} a_i x_i$. Multiplying (11.2.3) on both sides by a_{n+1} leads to

$$s_{n+1} - s_n \leq Ca_{n+1} + Ls_n a_{n+1}.$$

That is,

$$s_{n+1} \leq Ca_{n+1} + s_n(1 + La_{n+1}).$$

Iterating (with $x_0 \leq C$ by convention: replace C by $x_0 \vee C$ otherwise) we obtain

$$
\begin{aligned}
s_n &\leq C\sum_{k=0}^{n} \Pi_{m=k+1}^{n}(1 + La_m)a_k \\
&\leq C\int_0^{T_n} e^{L(T_n - s)}ds \\
&= \frac{C}{L}\left(e^{LT_n} - 1\right).
\end{aligned}
$$

(Here by convention, $\Pi_{m=n+1}^{n}(1 + La_m) = 1$.) Thus

$$x_{n+1} \leq C + Ls_n \leq C + L \times \frac{C}{L}\left(e^{LT_n} - 1\right) = Ce^{LT_n}. \qquad \blacksquare$$

11.2.2 Linear systems

A special and important class of differential equations is that of linear systems, i.e., the equations

$$\dot{x}(t) = A(t)x(t), \ t \geq t_0, x(t_0) = \bar{x}, \tag{11.2.4}$$

where $A(\cdot)$ is an $\mathcal{R}^{d \times d}$-valued continuous function of time. Although the right-hand side is now time-dependent, similar arguments to those of the preceding section establish its existence, uniqueness and continuous dependence on initial condition $\bar{x}$ *and* initial time t_0. It is easily seen that a linear combination of solutions of (11.2.4) will also be a solution for the corresponding linear combination of the initial conditions. Thus for given t_0, all solutions can be specified as linear combinations of solutions corresponding to $\bar{x} = e_j, 1 \leq j \leq d$, where $e_j \overset{\text{def}}{=}$ the unit vector in the jth coordinate direction. Let $\Phi(t, t_0)$ denote the $d \times d$ matrix whose ith column is $x(t)$ corresponding to $x(t_0) = e_i$ for $1 \leq i \leq d$. Then we have:

(i) $\Phi(t, t) = I_d$, the d-dimensional identity matrix,

(ii) $\Phi(t, s)\Phi(s, u) = \Phi(t, u)$ for $s, t, u \in \mathcal{R}$,

(iii) for $x(t_0) = \bar{x}$ in (11.2.4), $x(t) = \Phi(t, t_0)\bar{x}$.

For constant $A(\cdot) \equiv \bar{A}$, $\Phi(t, t_0) = \exp(\bar{A}(t - t_0))$, where the matrix exponential is defined by

$$e^{\bar{A}t} \overset{\text{def}}{=} \sum_{m=0}^{\infty} \frac{(\bar{A})^m t^m}{m!}.$$

An important instance of a linear system that we shall encounter is the following: Suppose $h(\cdot)$ in (11.2.1) is continuously differentiable with $Dh(x)$ denoting its Jacobian matrix evaluated at x. Then it can be shown that the map $\bar{x} \to x(t)$ for each t is continuously differentiable. If we denote by $Dx(t)$ its Jacobian matrix, then $Dx(\cdot)$ can be shown to satisfy the (matrix) linear system

$$\frac{d}{dt}Dx(t) = Dh(x(t))Dx(t), \ Dx(0) = I_d.$$

Formally, this may be derived simply by differentiating (11.2.1) on both sides w.r.t. the components of $\bar{x}$. In this case, $\{\Psi_t\}$ defined above is a flow of C^1-*diffeomorphisms*, i.e., continuously differentiable bijections with continuously differentiable inverses. This can be repeated for higher derivatives if sufficient regularity of h is available.

11.2.3 Asymptotic behaviour

Given a trajectory $x(\cdot)$ of (11.2.1), the set $\Omega \overset{\text{def}}{=} \cap_{t>0}\overline{\{x(s) : s > t\}}$, i.e., the set of its limit points as $t \to \infty$, is called its ω-limit set. (A similar definition for $t \to -\infty$ defines the 'α-limit set'.) In general this set depends upon the initial condition $\bar{x}$. Recall that a set A is positively (resp. negatively) invariant for (11.2.1) if $\bar{x} \in A$ implies that the corresponding $x(t)$ given by (11.2.1) is also in A for $t > 0$ (resp. $t < 0$), and is invariant if it is both positively and negatively invariant. It is easy to verify that Ω will be invariant. If $\Omega = \{x^*\}$, $x(t) \equiv x^*$ must be a trajectory of the o.d.e., whence $h(x^*) = 0$. Conversely, $h(x^*) = 0$

implies that $x(t) \equiv x^*$ defines a trajectory of the o.d.e., corresponding to $\bar{x} = x^*$ in (11.2.1). Such x^* are called equilibrium points of the o.d.e.

A compact (more generally, closed) invariant set M will be called an *attractor* if it has an open neighbourhood O such that every trajectory in O remains in O and converges to M. The largest such O is called the domain of attraction of M. A compact invariant set M will be said to be Liapunov stable if for any $\epsilon > 0$, there exists a $\delta > 0$ such that every trajectory initiated in the δ-neighbourhood of M remains in its ϵ-neighbourhood. A compact invariant set M is said to be asymptotically stable if it is both Liapunov stable and an attractor. If this $M = \{x^*\}$, the equilibrium point x^* is said to be asymptotically stable. One criterion for verifying asymptotic stability of x^* is 'Liapunov's second method': Suppose one can find a continuously differentiable function $V(\cdot)$ defined in a neighbourhood O of x^* such that $\langle \nabla V(x), h(x) \rangle < 0$ for $x^* \neq x \in O$ and $= 0$ for $x = x^*$, with $V(x) \to \infty$ as $x \to \partial O$ ($\overset{\text{def}}{=}$ the boundary of O). Then asymptotic stability of x^* follows from the observation that for any trajectory $x(\cdot)$ in O, $\frac{d}{dt}V(x(t)) \leq 0$ with equality only for $x(t) = x^*$. Conversely, asymptotic stability of x^* implies the existence of such a function (see, e.g., Krasovskii, 1963). This also generalizes to compact invariant sets M that are asymptotically stable.

If x^* is asymptotically stable and *all* trajectories of the o.d.e. converge to it, it is said to be *globally asymptotically stable*. In this case, O above may be taken to be the whole space. More generally, if one has a continuously differentiable $V : \mathcal{R}^d \to \mathcal{R}$ with $V(x) \to \infty$ as $\|x\| \to \infty$ and $\langle \nabla V(x), h(x) \rangle \leq 0 \; \forall x$, then any trajectory $x(\cdot)$ must converge to the largest invariant set contained in $\{x : \langle \nabla V(x), h(x) \rangle = 0\}$. This is known as the *LaSalle invariance principle*.

Not every equilibrium point need be asymptotically stable. This fact is best illustrated in the case of the constant coefficient linear system

$$\dot{x}(t) = Ax(t), \tag{11.2.5}$$

where A is a $d \times d$ matrix. We shall consider the case where all eigenvalues of A have nonzero real parts. This situation is 'structurally stable', i.e., invariant under small perturbations of A. In particular, A is nonsingular and thus the origin is the only equilibrium point. One can explicitly solve (11.2.5) as $x(t) = \exp(At)x(0)$. If all eigenvalues of A have strictly negative real parts, then $x(t) \to$ the origin exponentially. If not, it will do so only for those $x(0)$ that lie on the 'stable subspace', i.e., the eigenspace of those eigenvalues (if any) which have strictly negative real parts. It moves away from the origin eventually for any other initial condition, i.e., 'generically', meaning 'for initial conditions belonging to an open dense set'. This is because the stable subspace has codimension at least one by hypothesis and hence its complement is dense.

More generally, if h in (11.2.1) is continuously differentiable with Jacobian matrix $Dh(x^*)$ at an equilibrium point x^*, we may compare it with the linear system

$$\frac{d}{dt}(y(t) - x^*) = Dh(x^*)(y(t) - x^*), \qquad (11.2.6)$$

in a neighbourhood of x^*. If the eigenvalues of $Dh(x^*)$ have nonzero real parts, x^* is said to be a *hyperbolic* equilibrium point. Note that x^* is the unique equilibrium point for (11.2.6) in this case. It is known that in a small neighbourhood of a hyperbolic x^*, there exists a homeomorphism that maps trajectories of (11.2.5) with $A = Dh(x^*)$ and those of (11.2.6) into each other preserving orientation. Thus their qualitative behaviour is the same. (This is the *Hartman–Grossman theorem*.) Thus x^* is asymptotically stable for (11.2.1) if it is so for (11.2.6), i.e., if all eigenvalues of $Dh(x^*)$ have negative real parts. If x^* is not asymptotically stable for (11.2.1), then in a small neighbourhood of x^*, there exists a 'stable manifold' of dimension equal to the number of eigenvalues (if any) with negative real parts, such that if $x(0)$ lies on this manifold, $x(t) \to x^*$ and not otherwise.

Finally, we shall say that a probability measure μ on $\mathcal{R}^d$ is invariant under the flow $\{\Psi_t\}$ defined above if

$$\int f d\mu = \int f \circ \Psi_t d\mu \ \ \forall f \in C_b(\mathcal{R}^d) \text{ and } t \geq 0.$$

Define empirical measures $\nu(t), t \geq 0$, by

$$\int f d\nu(t) \stackrel{\text{def}}{=} \frac{1}{t} \int_0^t f(x(s)) ds, \ f \in C_b(\mathcal{R}^d), t \geq 0,$$

for $x(\cdot)$ as in (11.2.1). If $x(t)$ remains bounded as $t \uparrow \infty$, then $\{\nu(t)\}$ are supported on a compact set and hence are relatively compact in the space $\mathcal{P}(\mathcal{R}^d)$ of probability measures on $\mathcal{R}^d$ introduced in Appendix C. (This is a consequence of Prohorov's theorem mentioned in Appendix C.) Then every limit point of $\{\nu(t)\}$ as $t \to \infty$ is invariant under $\{\Psi_t\}$, as can be easily verified.

11.3 Appendix C: Topics in probability

11.3.1 Martingales

Let $(\Omega, \mathcal{F}, P)$ be a probability space and $\{\mathcal{F}_n\}$ a family of increasing sub-σ-fields of $\mathcal{F}$. A real-valued random process $\{X_n\}$ defined on this probability space is said to be a martingale w.r.t. the family $\{\mathcal{F}_n\}$ (or an $\{\mathcal{F}_n\}$-martingale) if it is integrable and

(i) X_n is $\mathcal{F}_n$-measurable for all n, and
(ii) $E[X_{n+1}|\mathcal{F}_n] = X_n$ a.s. for all n.

Alternatively, one says that $(X_n, \mathcal{F}_n)_{n \geq 0}$ is a martingale. The sequence $M_n = X_n - X_{n-1}$ is then called a martingale difference sequence. The reference to $\{\mathcal{F}_n\}$ is often dropped when it is clear from the context. A sequence of $\mathcal{R}^d$-valued random variables is said to be a (vector) martingale if each of its component processes is. There is a very rich theory of martingales and related processes such as submartingales (in which '=' is replaced by '$\geq$' in (ii) above), supermartingales (in which '=' is replaced by '$\leq$' in (ii) above), 'almost supermartingales', and so on. (In fact, the purely probabilistic approach to stochastic approximation relies heavily on these.) We shall confine ourselves to listing a few key facts that have been used in this book. For more, the reader may refer to Borkar (1995), Breiman (1968), Neveu (1970) or Williams (1991). The results presented here for which no specific reference is given will be found in particular in Borkar (1995). Throughout what follows, $\{X_n\}, \{\mathcal{F}_n\}$ are as above.

(i) *A decomposition theorem*

Theorem 9. *Let $\{M_n\}$ be a d-dimensional $\{\mathcal{F}_n\}$-martingale such that $E[\|M_n\|^2] < \infty \; \forall n$. Then there exists an $\mathcal{R}^{d \times d}$-valued process $\{\Gamma_n\}$ such that Γ_n is $\mathcal{F}_{n-1}$-measurable for all n and $\{M_n M_n^T - \Gamma_n\}$ is an $\mathcal{R}^{d \times d}$-valued $\{\mathcal{F}_n\}$-martingale.*

This theorem is just a special case of the *Doob decomposition*. In fact, it is easy to see that for

$$M_n = [M_n(1), \ldots, M_n(d)]^T,$$

one has $\Gamma_n = [[\Gamma_n(i,j)]]_{1 \leq i,j \leq d}$, with

$$\Gamma_n(i,j) = \sum_{m=1}^{n} E[M_m(i)M_m(j) - M_{m-1}(i)M_{m-1}(j)|\mathcal{F}_{m-1}]$$

for $1 \leq i, j \leq d$.

(ii) *Convergence theorems*

Theorem 10. *If $\sup_n E[X_n^+] < \infty$, then $\{X_n\}$ converges a.s.*

Theorem 11. *If $E[X_n^2] < \infty \; \forall n$, then $\{X_n\}$ converges a.s. on the set $\{\sum_n E[(X_{n+1} - X_n)^2|\mathcal{F}_n] < \infty\}$ and is $o(\sum_{m=1}^{n-1} E[(X_{m+1} - X_m)^2|\mathcal{F}_m])$ a.s. on the set $\{\sum_n E[(X_{n+1} - X_n)^2|\mathcal{F}_n] = \infty\}$.*

The following result is sometimes useful:

Theorem 12. *If $E[\sup_n |X_{n+1} - X_n|] < \infty$, then*

$$P(\{\{X_n\} \; converges\} \cup \{\limsup_{n \to \infty} X_n = -\liminf_{n \to \infty} X_n = \infty\}) = 1.$$

(iii) *Inequalities:* Let $E[|X_n|^p] < \infty \; \forall n$ for some $p \in (1, \infty)$. Suppose $X_0 = 0$, implying in particular that $E[X_n] = 0 \; \forall n$.

Theorem 13. *(Burkholder inequality) There exist constants $c, C > 0$ depending on p alone such that for all $n = 1, 2, \ldots, \infty$,*

$$cE[(\sum_{m=1}^{n} (X_m - X_{m-1})^2)^{\frac{p}{2}}] \leq E[\sup_{m \leq n} |X_m|^p]$$

$$\leq CE[(\sum_{m=1}^{n} (X_m - X_{m-1})^2)^{\frac{p}{2}}].$$

Theorem 14. *(Concentration inequality) Suppose that*

$$|X_n - X_{n-1}| \leq k_n < \infty$$

for some deterministic constants $\{k_n\}$. Then for $\lambda > 0$,

$$P(\sup_{m \leq n} |X_m| > \lambda) \leq 2e^{-\frac{\lambda^2}{\sum_{m \leq n} k_m^2}}.$$

See McDiarmid (1998, p. 227) for a proof of this result.

(iv) *Central limit theorem:* We shall state a more general 'central limit theorem for vector martingale arrays'. For $n \geq 1$, let $(M_m^n, \mathcal{F}_m^n), m \geq 0$, be $\mathcal{R}^d$-valued vector martingales with $E[||M_m^n||^2] < \infty \; \forall m, n$. Define $\{\Gamma_m^n\}$ as above (i.e., Γ_m^n is $\mathcal{F}_{m-1}^n$-measurable for all n, m, and $M_m^n (M_m^n)^{\mathrm{T}} - \Gamma_m^n, m \geq 1$, is an $\{\mathcal{F}_m^n\}$-martingale for each n). Recall that a stopping time with respect to the increasing family of σ-fields $\{\mathcal{F}_n\}$ is a random variable taking values in $\{0, 1, \ldots, \infty\}$ such that for all n in this set, $\{\tau \leq n\}$ is $\mathcal{F}_n$-measurable (with $\mathcal{F}_\infty \stackrel{\mathrm{def}}{=} \vee_n \mathcal{F}_n$).

Theorem 15. *(Central limit theorem) Suppose that there exist $\{\mathcal{F}_m^n\}$-stopping times τ_n for $n \geq 1$ such that $\tau_n \uparrow \infty$ a.s. and:*

(a) *for some symmetric positive definite $\Gamma \in \mathcal{R}^{d \times d}$, $\Gamma_{\tau_n}^n \to \Gamma$ in probability, and*

(b) *for any $\epsilon > 0$,*

$$\sum_{m=1}^{\tau_n} E[||M_m^n - M_{m-1}^n||^2 I\{||M_m^n - M_{m-1}^n|| > \epsilon\} | \mathcal{F}_{m-1}^n] \to 0$$

in probability.

Then $\{M_{\tau_n}^n\}$ converges in law to the d-dimensional Gaussian measure with zero mean and covariance matrix Γ.

See Hall and Heyde (1980) for this and related results. (The vector case stated here follows from the scalar case on considering arbitrary one dimensional projections.)

11.3.2 Spaces of probability measures

Let S be a metric space with a complete metric $d(\cdot, \cdot)$. Endow S with its Borel σ-field, i.e., the σ-field generated by the open d-balls. Assume also that S is separable, i.e., has a countable dense subset $\{s_n\}$. Let $\mathcal{P}(S)$ denote the space of probability measures on S. $\mathcal{P}(S)$ may be metrized with the metric

$$\rho(\mu, \nu) \stackrel{\text{def}}{=} \inf E[d(X, Y) \wedge 1],$$

where the infimum is over all pairs of S-valued random variables X, Y such that the law of X is μ and the law of Y is ν. This metric can be shown to be complete. Let δ_x denote the Dirac measure at $x \in S$, i.e., $\delta_x(A) = 0$ or 1 depending on whether $x \notin A$ or $\in A$ for A Borel in S. Also, let $\{s_i, i \geq 1\}$ denote a prescribed countable dense subset of S. Then $\mu \in \mathcal{P}(S)$ of the form

$$\mu = \sum_{k=1}^{m} a_k \delta_{x_k},$$

for some $m \geq 1$, $a_1, \ldots, a_m$ rational in $[0, 1]$ with $\sum_i a_i = 1$, and $\{x_i, 1 \leq i \leq m\} \subset \{s_i, i \geq 1\}$, are countable dense in $\mathcal{P}(S)$. Hence $\mathcal{P}(S)$ is separable. The following theorem gives several equivalent formulations of convergence in $\mathcal{P}(S)$:

Theorem 16. *The following are equivalent:*

 (i) $\rho(\mu_n, \mu) \to 0$.
 (ii) *For all $f \in C_b(S)$, $\int f d\mu_n \to \int f d\mu$.*
 (iii) *For all $f \in C_b(S)$ that are uniformly continuous w.r.t. some compatible metric on S, $\int f d\mu_n \to \int f d\mu$.*
 (iv) *For all open $G \subset S$, $\liminf_{n \to \infty} \mu_n(G) \geq \mu(G)$.*
 (v) *For all closed $F \subset S$, $\limsup_{n \to \infty} \mu_n(F) \leq \mu(F)$.*
 (vi) *For all $A \subset S$ satisfying $\mu(\partial A) = 0$, $\mu_n(A) \to \mu(A)$.*

In fact, there exists a countable set $\{f_i, i \geq 1\} \subset C_b(S)$ such that $\rho(\mu_n, \mu) \to 0$ if and only if $\int f_i d\mu_n \to \int f_i d\mu$ $\forall i$. This set is known as a convergence determining class. If S is compact, $C(S)$ is separable and any countable dense set in its unit ball will do. For non-compact S, embed it densely and homeomorphically into a compact subset $\bar{S}$ of $[0, 1]^\infty$, consider a countable subset of $C(\bar{S})$, and restrict it to S (see Borkar, 1995, Chapter 2).

Relative compactness in $\mathcal{P}(S)$ is characterized by the following theorem: Say that $\mathcal{A} \subset \mathcal{P}(S)$ is a tight set if for any $\epsilon > 0$, there exists a compact $K_\epsilon \subset S$ such that

$$\mu(K_\epsilon) > 1 - \epsilon \ \forall \mu \in \mathcal{A}.$$

By a result of Oxtoby and Ulam, every singleton in $\mathcal{P}(S)$ is tight (see, e.g., Borkar (1995), p. 4).

Theorem 17. *(Prohorov)* $A \subset \mathcal{P}(S)$ *is relatively compact if and only if it is tight.*

The following theorem is extremely important:

Theorem 18. *(Skorohod)* *If $\mu_n \to \mu_\infty$ in $\mathcal{P}(S)$, then on some probability space there exist random variables $X_n, n = 1, 2, \ldots, \infty$, such that the law of X_n is μ_n for each $n, 1 \le n \le \infty$, and $X_n \to X_\infty$ a.s.*

A stronger convergence notion than convergence in $\mathcal{P}(S)$ is that of convergence in total variation. We say that $\mu_n \to \mu$ in total variation if

$$\sup \left| \int f d\mu_n - \int f d\mu \right| \to 0,$$

where the supremum is over all $f \in C_b(S)$ with $\sup_x |f(x)| \le 1$. This in turn allows us to write $\int f d\mu_n \to \int f d\mu$ for *bounded measurable* $f : S \to \mathcal{R}$. The next theorem gives a useful test for convergence in total variation. We shall say that $\mu \in \mathcal{P}(S)$ is *absolutely continuous* with respect to a positive, not necessarily finite measure λ on S if $\lambda(B) = 0$ implies $\mu(B) = 0$ for any Borel $B \subset S$. It is known that in this case there exists a measurable $\Lambda : S \to \mathcal{R}$ such that $\int f d\mu = \int f \Lambda d\lambda$ for all μ-integrable $f : S \to \mathcal{R}$. This is the *Radon–Nikodym theorem* of measure theory and $\Lambda(\cdot)$ is called the *Radon–Nikodym derivative* of μ w.r.t. λ. For example, the familiar probability density is the Radon–Nikodym derivative of the corresponding probability measure w.r.t. the Lebesgue measure. The likelihood ratio in statistics is another example of a Radon–Nikodym derivative.

Theorem 19. *(Scheffé)* *Suppose $\mu_n, n = 1, 2, \ldots, \infty$, are absolutely continuous w.r.t. a positive measure λ on S with Λ_n the corresponding Radon–Nikodym derivatives. If $\Lambda_n \to \Lambda_\infty$ λ-a.s., then $\mu_n \to \mu_\infty$ in total variation.*

11.3.3 Stochastic differential equations

As this topic has been used only nominally and that too in only one place, viz., Chapter 8, we shall give only the barest facts. An interested reader can find much more in standard texts such as Oksendal (2005). Consider a probability space $(\Omega, \mathcal{F}, P)$ with a family $\{\mathcal{F}_t, t \ge 0\}$ of sub-σ-fields of $\mathcal{F}$ satisfying:

(i) it is increasing, i.e., $\mathcal{F}_s \subset \mathcal{F}_t \ \forall t > s$,

(ii) it is right continuous, i.e., $\mathcal{F}_t = \cap_{s>t} \mathcal{F}_s \ \forall t$,

(iii) it is complete, i.e., each $\mathcal{F}_t$ contains all zero probability sets in $\mathcal{F}$ and their subsets.

A measurable stochastic process $\{Z_t\}$ on this probability space is said to be adapted to $\{\mathcal{F}_t\}$ if Z_t is $\mathcal{F}_t$-measurable $\forall t \ge 0$. A d-dimensional Brownian motion $W(\cdot)$ defined on $(\Omega, \mathcal{F}, P)$ is said to be an $\{\mathcal{F}_t\}$-Wiener process in $\mathcal{R}^d$ if it is adapted to $\{\mathcal{F}_t\}$ and for each $t \geq 0$, $W(t + \cdot) - W(t)$ is independent of $\mathcal{F}_t$. Let $\{\xi_t\}$ be an $\mathcal{R}^d$-valued process satisfying:

(i) it is adapted to $\{\mathcal{F}_t\}$,

(ii) $E[\int_0^t \|\xi_s\|^2 ds] < \infty \ \forall t > 0$,

(iii) there exist $0 = t_0 < t_1 < t_2 < \cdots < t_i \overset{i \uparrow \infty}{\to} \infty$ such that $\xi_t =$ some $\mathcal{F}_{t_i}$-measurable random variable ζ_i for $t \in [t_i, t_{i+1})$ (i.e., $\{\xi_t\}$ is a piecewise constant adapted process).

Define the stochastic integral of $\{\xi_t\}$ with respect to the $\{\mathcal{F}_t\}$-Wiener process $W(\cdot)$ by

$$\int_0^t \langle \xi_s, dW(s) \rangle \overset{\text{def}}{=} \sum_{0 < i \leq i^*(t)} \langle \zeta_{i-1}, (W(t_i) - W(t_{i-1})) \rangle$$
$$+ \langle \zeta_{i^*(t)}, (W(t) - W(t_{i^*(t)})) \rangle,$$

where $i^*(t)$ is the unique integer $i \geq 0$ such that $t \in [t_{i^*(t)}, t_{i^*(t)+1})$. From the 'independent increments' property of Brownian motion, one can verify that

$$E[|\int_0^t \langle \xi_s, dW(s) \rangle|^2] = E[\int_0^t \|\xi_s\|^2 ds] \ \forall t > 0.$$

More generally, let $\{\xi_t\}$ be an $\mathcal{R}^d$-valued process satisfying (i)-(ii) above and let $\{\xi_t^n\}, n \geq 1$, be a family of real-valued processes satisfying (i)-(iii) above, such that

$$E[\int_0^t \|\xi_s - \xi_s^n\|^2 ds] \to 0 \ \forall t > 0. \tag{11.3.1}$$

Then once again using the independent increments property of the Wiener process $W(\cdot)$, we have, in view of (11.3.1),

$$E[|\int_0^t \langle \xi_s^n, dW(s) \rangle - \int_0^t \langle \xi_s^m, dW(s) \rangle|^2] = E[\int_0^t \|\xi_s^n - \xi_s^m\|^2 ds] \to 0 \ \forall t > 0.$$

That is, the sequence of random variables $\int_0^t \langle \xi_s^n, dW(s) \rangle, n \geq 1$, is Cauchy in $L_2(\Omega, \mathcal{F}, P)$ and hence has a unique limit therein, which we denote by $\int_0^t \langle \xi_s, dW(s) \rangle$.

Next we argue that it is always possible to find such $\{\xi_t^n\}, n \geq 1$, for any $\{\xi_t\}$ satisfying (i)-(ii). Here's a sketch of what is involved: For a small $a > 0$, define $\{\xi_t^{(a)}\}$ by

$$\xi_t^{(a)} = \frac{1}{a \wedge t} \int_{(t-a) \vee 0}^t \xi_s ds, \ t \geq 0,$$

which is adapted (i.e., $\mathcal{F}$-measurable for each t), has continuous paths, and approximates $\{\xi_s\}$ on $[0, t]$ in mean square to any desired accuracy for a small enough. Now pick a 'grid' $\{t_i\}$ as above and define

$$\tilde{\xi}_s = \xi_{t_i}^{(a)}, \; s \in [t_i, t_{i+1}), \; i \geq 0.$$

This can approximate $\{\xi_s^{(a)}\}$ arbitrarily closely in mean square if the grid is taken to be sufficiently fine.

Although this construction was for a fixed $t > 0$, one can show that it is possible to construct this process so that $t \to \int_0^t \xi_s dW(s)$ is continuous in t a.s. We call this the stochastic integral of $\{\xi_t\}$ w.r.t. $W(\cdot)$.

Let $m : \mathcal{R}^d \times [0, \infty) \to \mathcal{R}^d$ and $\sigma : \mathcal{R}^d \times [0, \infty) \to \mathcal{R}^{d \times d}$ be Lipschitz maps. Consider the stochastic integral equation

$$X(t) = X_0 + \int_0^t m(X(s), s) ds + \int_0^t \langle \sigma(X(s), s), dW(s) \rangle, \; t \geq 0, \quad (11.3.2)$$

where X_0 is an $\mathcal{F}_0$-measurable random variable. It is standard practice to call (11.3.2) a stochastic *differential* equation and write it as

$$dX(t) = m(X(t), t) dt + \sigma(X(t), t) dW(t), \; X(0) = X_0.$$

It is possible to show that this will have an a.s. unique solution $X(\cdot)$ on $(\Omega, \mathcal{F}, P)$ with continuous paths. (This is the so-called *strong* solution. There is also a notion of a *weak* solution, which we shall not concern ourselves with.) Clearly, the case of linear or constant $m(\cdot)$ and $\sigma(\cdot)$ is covered by this.

The equation

$$dX(t) = A(t) X(t) dt + D(t) dW(t)$$

with Gaussian $X(0) = X_0$ is a special case of the above. $X(\cdot)$ is then a Gaussian and Markov process. This equation can be explicitly 'integrated' as follows: Let $\Phi(t, t_0), t \geq t_0$, be the unique solution to the linear matrix differential equation

$$\frac{d}{dt}\Phi(t, t_0) = A(t)\Phi(t, t_0), \; t \geq t_0; \; \Phi(t_0, t_0) = I_d,$$

where $I_d \in \mathcal{R}^{d \times d}$ is the identity matrix. (See Appendix B. Recall in particular that for $A(\cdot) \equiv$ a constant matrix $\bar{A}$, $\Phi(t, t_0) = \exp(\bar{A}(t - t_0))$.) Then

$$X(t) = \Phi(t, t_0) X_0 + \int_0^t \langle \Phi(t, s), D(s) dW(s) \rangle, \; t \geq 0.$$

Both the Gaussian property (when X_0 is Gaussian) and the Markov property of $X(\cdot)$ can be easily deduced from this using the Gaussian and independent increments properties of $W(\cdot)$.

References

[1] ABOUNADI, J.; BERTSEKAS, D. P.; BORKAR, V. S. (2001) 'Learning algorithms for Markov decision processes with average cost', *SIAM Journal on Control and Optimization* 40, 681–698.

[2] ABOUNADI, J.; BERTSEKAS, D. P.; BORKAR, V. S. (2002) 'Stochastic approximation for nonexpansive maps: applications to Q-learning algorithms', *SIAM Journal on Control and Optimization* 41, 1–22.

[3] ANBAR, D. (1978) 'A stochastic Newton-Raphson method', *Journal of Statistical Planning and Inference* 2, 153–163.

[4] ARTHUR, W. B. (1994) *Increasing Returns and Path Dependence in the Economy*, Univ. of Michigan Press, Ann Arbor, Mich.

[5] ARTHUR, W. B.; ERMOLIEV, Y.; KANIOVSKI, Y. (1983) 'A generalized urn problem and its applications', *Cybernetics* 19, 61–71.

[6] AUBIN, J. P.; CELLINA, A. (1984) *Differential Inclusions*, Springer Verlag, Berlin.

[7] AUBIN, J. P.; FRANKOWSKA, H. (1990) *Set-Valued Analysis*, Birkhäuser, Boston.

[8] BALAKRISHNAN, A. V. (1976) *Applied Functional Analysis*, Springer Verlag, New York.

[9] BARDI, M.; CAPUZZO-DOLCETTA, I. (1997) *Optimal Control and Viscosity Solutions of Hamilton-Jacobi-Bellman Equations*, Birkhäuser, Boston.

[10] BARTO, A., SUTTON, R.; ANDERSON, C. (1983) 'Neuron-like elements that can solve difficult learning control problems', *IEEE Transactions on Systems, Man and Cybernetics*, 13, 835–846.

[11] BENAIM, M. (1996) 'A dynamical system approach to stochastic approximation', *SIAM Journal on Control and Optimization* 34, 437–472.

[12] BENAIM, M. (1997) 'Vertex-reinforced random walks and a conjecture of Pemantle', *Annals of Probability* 25, 361–392.

[13] BENAIM, M. (1999) 'Dynamics of stochastic approximation algorithms', in *Le Séminaire de Probabilités*, J. Azéma, M. Emery, M. Ledoux and M. Yor (eds.), Springer Lecture Notes in Mathematics No. 1709, Springer Verlag, Berlin-Heidelberg, 1–68.

[14] BENAIM, M.; HIRSCH, M. (1997) 'Stochastic adaptive behaviour for prisoner's dilemma', *preprint*.

[15] BENAIM, M.; HOFBAUER, J.; SORIN, S. (2005) 'Stochastic approximation and differential inclusions', *SIAM Journal on Control and Optimization* 44, 328-348.

[16] BENAIM, M.; SCHREIBER, S. (2000) 'Ergodic properties of weak asymptotic pseudotrajectories for semiflows', *Journal of Dynamics and Differential Equations* 12, 579–598.

[17] BENVENISTE, A.; METIVIER, M.; PRIOURET, P. (1990) *Adaptive Algorithms*

and *Stochastic Approximation*, Springer Verlag, Berlin - New York.

[18] BERTSEKAS, D. P.; TSITSIKLIS, J. N. (1996) *Neuro-Dynamic Programming*, Athena Scientific, Belmont, Mass.

[19] BHATNAGAR, S.; BORKAR, V. S. (1997) 'Multiscale stochastic approximation for parametric optimization of hidden Markov models', *Probability in the Engineering and Informational Sciences*, 11, 509–522.

[20] BHATNAGAR, S.; BORKAR, V. S. (1998) 'A two time-scale stochastic approximation scheme for simulation-based parametric optimization', *Probability in the Engineering and Informational Sciences*, 12, 519–531.

[21] BHATNAGAR, S.; FU, M. C.; MARCUS, S. I., WANG, I.-J. (2003) 'Two-timescale simultaneous perturbation stochastic approximation using deterministic perturbation sequences', *ACM Transactions on Modelling and Computer Simulation*, 13, 180–209.

[22] BORDER, K. C. (1989) *Fixed Point Theorems with Applications to Economics and Game Theory*, Cambridge Univ. Press, Cambridge, UK.

[23] BORKAR, V. S. (1995) *Probability Theory: An Advanced Course*, Springer Verlag, New York.

[24] BORKAR, V. S. (1997) 'Stochastic approximation with two time scales', *Systems and Control Letters* 29, 291–294.

[25] BORKAR V. S. (1998) 'Asynchronous stochastic approximation', *SIAM Journal on Control and Optimization* 36, 840–851 (Correction note in *ibid.*, 38, 662–663).

[26] BORKAR, V. S. (2002) 'On the lock-in probability of stochastic approximation', *Combinatorics, Probability and Computing* 11, 11–20.

[27] BORKAR, V. S. (2003) 'Avoidance of traps in stochastic approximation', *Systems and Control Letters* 50, 1–9 (Correction note in *ibid.* (2006) 55, 174–175).

[28] BORKAR, V. S. (2005) 'An actor-critic algorithm for constrained Markov decision processes', *Systems and Control Letters* 54, 207–213.

[29] BORKAR, V. S. (2006) 'Stochastic approximation with 'controlled Markov' noise', *Systems and Control Letters* 55, 139–145.

[30] BORKAR, V. S.; KUMAR, P. R. (2003) 'Dynamic Cesaro-Wardrop equilibration in networks', *IEEE Transactions on Automatic Control* 48, 382–396.

[31] BORKAR, V. S.; MANJUNATH, D. (2004) 'Charge based control of diffserve-like queues', *Automatica* 40, 2040–2057.

[32] BORKAR, V. S.; MEYN, S. P. (2000) 'The O.D.E. method for convergence of stochastic approximation and reinforcement learning', *SIAM Journal on Control and Optimization* 38, 447–469.

[33] BORKAR, V. S.; SOUMYANATH, K. (1997) 'A new analog parallel scheme for fixed point computation, Part 1: Theory', *IEEE Transactions on Circuits and Systems I: Fundamental Theory and Applications*, 44, 351–355.

[34] BRANDIÈRE, O. (1998) 'Some pathological traps for stochastic approximation', *SIAM Journal on Control and Optimization* 36, 1293–1314.

[35] BRANDIÈRE, O.; DUFLO, M. (1996) 'Les algorithmes stochastiques contournent – ils les pieges?', *Annales de l'Institut Henri Poincaré* 32, 395–427.

[36] BREIMAN, L. (1968) *Probability*, Addison-Wesley, Reading, Mass.

[37] BROWN, G.; (1951) 'Iterative solutions of games with fictitious play', in *Activity Analysis of Production and Allocation*, T. Koopmans (ed.), John Wiley, New York.

[38] CHEN, H.-F. (1994) 'Stochastic approximation and its new applications', in *Proc. 1994 Hong Kong International Workshop on New Directions in Control and Manufacturing*, 2–12.

[39] CHEN, H.-F. (2002) *Stochastic Approximation and Its Applications*, Kluwer Academic, Dordrecht, The Netherlands.

[40] CHOW, Y. S.; TEICHER, H. (2003) *Probability Theory: Independence, Inter-*

changeability, Martingales (3rd ed.), Springer Verlag, New York.

[41] CHUNG, K. L. (1954) 'On a stochastic approximation method', *Annals of Mathematical Statistics* 25, 463–483.

[42] CUCKER, F.; SMALE, S. (2007) 'Emergent behavior in flocks', *IEEE Transactions on Automatic Control* 52, 852–862.

[43] DEREVITSKII, D. P.; FRADKOV, A. L. (1974) 'Two models for analyzing the dynamics of adaptation algorithms', *Automation and Remote Control* 35, 59–67.

[44] DUFLO, M. (1996) *Algorithmes Stochastiques*, Springer Verlag, Berlin-Heidelberg.

[45] DUFLO, M. (1997) *Random Iterative Models*, Springer Verlag, Berlin-Heidelberg.

[46] DUPUIS, P. (1988) 'Large deviations analysis of some recursive algorithms with state-dependent noise', *Annals of Probability* 16, 1509–1536.

[47] DUPUIS, P.; KUSHNER, H. J. (1989) 'Stochastic approximation and large deviations: upper bounds and w. p. 1 convergence', *SIAM Journal on Control and Optimization* 27, 1108–1135.

[48] DUPUIS, P.; NAGURNEY, A. (1993) 'Dynamical systems and variational inequalities, *Annals of Operations Research* 44, 7–42.

[49] FABIAN, V. (1960) 'Stochastic approximation methods', *Czechoslovak Mathematical Journal* 10, 125–159.

[50] FABIAN, V. (1968) 'On asymptotic normality in stochastic approximation', *Annals of Mathematical Statistics* 39, 1327–1332.

[51] FANG, H.-T.; CHEN, H.-F. (2000) 'Stability and instability of limit points for stochastic approximation algorithms', *IEEE Transactions on Automatic Control* 45, 413–420.

[52] FILIPPOV, A. F. (1988) *Differential Equations with Discontinuous Righthand Sides*, Kluwer Academic, Dordrecht.

[53] FORT, J.-C.; PAGES, G. (1995) 'On the a.s. convergence of the Kohonen algorithm with a general neighborhood function', *Annals of Applied Probability* 5, 1177–1216.

[54] FU, M. C.; HU, J.-Q. (1997) *Conditional Monte Carlo: Gradient Estimation and Optimization Applications*, Kluwer Academic, Boston.

[55] FUDENBERG, D.; LEVINE, D. (1998) *Theory of Learning in Games*, MIT Press, Cambridge, Mass.

[56] GELFAND, S. B.; MITTER, S. K. (1991) 'Recursive stochastic algorithms for global optimization in R^d', *SIAM Journal on Control and Optimization* 29, 999–1018.

[57] GERENCSÉR, L. (1992) 'Rate of convergence of recursive estimators', *SIAM Journal on Control and Optimization* 30, 1200–1227.

[58] GOLDSTEIN, L. (1988) 'On the choice of step-size in the Robbins-Monro procedure', *Statistics and Probability Letters* 6, 299–303.

[59] GLASSERMAN, P. (1991) *Gradient Estimation via Perturbation Analysis*, Kluwer Academic, Boston.

[60] GROSSBERG, S. (1978) 'Competition, decision and consensus', *Journal of Mathematical Analysis and Applications* 66, 470–493.

[61] HALL, P.; HEYDE, C. C. (1980) *Martingale Limit Theory and Its Applications*, Academic Press, New York.

[62] HARTMAN, P. (1982) *Ordinary Differential Equations* (2nd ed.), Birkhäuser, Boston.

[63] HAYKIN, S. (2001) *Adaptive Filter Theory* (4th ed.), Prentice Hall, Englewood Cliffs, N.J.

[64] HAYKIN, S. (1998) *Neural Networks: A Comprehensive Foundation* (2nd ed.), McMillan Publ. Co., New York.

[65] HELMKE, U.; MOORE, J. B (1994) *Optimization and Dynamical Systems*, Springer Verlag, London.

[66] HERTZ, J.; KROGH, A.; PALMER, R. (1991) *An Introduction to the Theory of Neural Computation*, Addison Wesley, Redwood City, Calif.

[67] HIRSCH, M. W. (1985) 'Systems of differential equations that are competitive or cooperative II: Convergence almost everywhere', *SIAM Journal on Mathematical Analysis* 16, 423–439.

[68] HIRSCH, M. W.; SMALE, S.; DEVANEY, R. (2003) *Differential Equations, Dynamical Systems and an Introduction to Chaos*, Academic Press, New York.

[69] HSIEH, M.-H.; GLYNN, P. W. (2002) 'Confidence regions for stochastic approximation algorithms', *Proc. of the Winter Simulation Conference* 1, 370–376.

[70] HO, Y. C.; CAO, X. (1991) *Perturbation Analysis of Discrete Event Dynamical Systems*, Birkhäuser, Boston.

[71] HOFBAUER, C.; SIEGMUND, K. (1998) *Evolutionary Games and Population Dynamics*, Cambridge Univ. Press, Cambridge, UK.

[72] HU, J.; WELLMAN, M. P. (1998) 'Multiagent reinforcement learning: theoretical framework and an algorithm', *Proc. of the 15th International Conference on Machine Learning*, Madison, Wisc., 242–250.

[73] JAFARI, A., GREENWALD, A.; GONDEK, D.; ERCAL, G. (2001) 'On no-regret learning, fictitious play and Nash equilibrium', *Proc. of the 18th International Conference on Machine Learning*, Williams College, Williamstown, Mass., 226–233.

[74] KAILATH, T. (1980) *Linear Systems*, Prentice Hall, Englewood Cliffs, N.J.

[75] KANIOVSKI, Y. M.; YOUNG, H. P. (1995) 'Learning dynamics in games with stochastic perturbations', *Games and Economic Behavior* 11, 330–363.

[76] KATKOVNIK, V.; KULCHITSKY, Y. (1972) 'Convergence of a class of random search algorithms', *Automation and Remote Control* 8, 1321–1326.

[77] KEIFER, J.; WOLFOWITZ, J. (1952) 'Stochastic estimation of the maximum of a regression function', *Annals of Mathematical Statistics* 23, 462–466.

[78] KELLY, F. P.; MAULLOO, A.; TAN, D. (1998) 'Rate control in communication networks: shadow prices, proportional fairness and stability', *Journal of Operational Research Society* 49, 237–252.

[79] KOHONEN, T. (2002) 'Learning vector quantization', in *The Handbook of Brain Theory and Neural Networks* (2nd ed.), M. A. Arbib (ed.), MIT Press, Cambridge, Mass., 537–540.

[80] KOSMOTOPOULOS, E. B.; CHRISTODOULOU, M. A. (1996) 'Convergence properties of a class of learning vector quantization algorithms', *IEEE Transactions on Image Processing* 5, 361–368.

[81] KRASOVSKII, N. N. (1963) *Stability of Motion*, Stanford Univ. Press, Stanford, Calif.

[82] KUSHNER, H. J.; CLARK, D. (1978) *Stochastic Approximation Algorithms for Constrained and Unconstrained Systems*, Springer Verlag, New York.

[83] KUSHNER, H. J.; YIN, G. (1987a) 'Asymptotic properties for distributed and communicating stochastic approximation algorithms', *SIAM Journal on Control and Optimization* 25, 1266–1290.

[84] KUSHNER, H. J.; YIN, G. (1987b) 'Stochastic approximation algorithms for parallel and distributed processing', *Stochastics and Stochastics Reports* 22, 219–250.

[85] KUSHNER, H. J.; YIN, G. (2003) *Stochastic Approximation and Recursive Algorithms and Applications* (2nd ed.), Springer Verlag, New York.

[86] LAI, T. L. (2003) 'Stochastic approximation', *Annals of Statistics* 31, 391–406.

[87] LAI, T. L.; ROBBINS, H. (1978) 'Limit theorems for weighted sums and stochastic approximation processes', *Proc. National Academy of Sciences USA* 75, 1068–1070.

[88] LEIZAROWITZ, A. (1997) 'Convergence of solutions to equations arising in neural networks', *Journal of Optimization Theory and Applications* 94, 533–560.

[89] LI, Y. (2003) 'A martingale inequality and large deviations', *Statistics and Probability Letters* 62, 317–321.

[90] LITTMAN, M. L. (2001) 'Value-function reinforcement learning in Markov games', *Cognitive Systems Research* 2, 55–66.

[91] LJUNG, L. (1977) 'Analysis of recursive stochastic algorithms', *IEEE Transactions on Automatic Control* 22, 551–575.

[92] LJUNG, L. (1978) 'Strong convergence of a stochastic approximation algorithm'. *Annals of Statistics* 6, 680–696.

[93] LJUNG, L. (1999) *System Identification: Theory for the User* (2nd ed.), Prentice Hall, Englewood Cliffs, N.J.

[94] LJUNG, L.; PFLUG, G. C.; WALK, H. (1992) *Stochastic Approximation and Optimization of Random Systems*, Birkhäuser, Basel.

[95] MATSUMOTO, Y. (2002) *An Introduction to Morse Theory*, Trans. of Mathematical Monographs No. 208, American Math. Society, Providence, R.I.

[96] McDIARMID, C. (1998) 'Concentration', in *Probabilistic Methods for Algorithmic Discrete Mathematics*, M. Habib, C. McDiarmid, J. Ramirez-Alfonsin and B. Reed (eds.), Springer Verlag, Berlin-Heidelberg.

[97] MEL'NIKOV, A. (1996) 'Stochastic differential equations: singularity of coefficients, regression models and stochastic approximation', *Russian Mathematical Surveys* 52, 819–909.

[98] MILGROM, P.; SEGAL, I (2002) 'Envelope theorems for arbitrary choice sets'. *Econometrica* 70, 583–601.

[99] NEVELSON, M.; KHASMINSKII, R. (1976) *Stochastic Approximation and Recursive Estimation*, Trans. of Mathematical Monographs No. 47, American Math. Society, Providence, R.I.

[100] NEVEU, J. (1975) *Discrete Parameter Martingales*, North Holland, Amsterdam.

[101] OJA, E. (1982) 'Simplified neuron model as a principal component analyzer', *Journal of Mathematical Biology* 15, 267–273.

[102] OJA, E.; KARHUNEN, J. (1985) 'On stochastic approximation of the eigenvectors and eigenvalues of the expectation of a random matrix', *Journal of Mathematical Analysis and Applications* 106, 69–84.

[103] OKSENDAL, B. (2005) *Stochastic Differential Equations* (6th ed.), Springer Verlag, Berlin-Heidelberg.

[104] ORTEGA, J. M.; RHEINBOLDT, W. C. (2000) *Iterative Solutions of Nonlinear Equations in Several Variables*, Society for Industrial and Applied Math.. Philadelphia.

[105] PELLETIER, M. (1998) 'On the almost sure asymptotic behaviour of stochastic algorithms', *Stochastic Processes and Their Applications* 78, 217–244.

[106] PELLETIER, M. (1999) 'An almost sure central limit theorem for stochastic approximation algorithms', *Journal of Multivariate Analysis* 71, 76–93.

[107] PEMANTLE, R. (1990) 'Nonconvergence to unstable points in urn models and stochastic approximations', *Annals of Probability* 18, 698–712.

[108] PEMANTLE, R. (2007) 'A survey of random processes with reinforcement', *Probability Surveys* 7, 1–79.

[109] PEZESHKI-ESFAHANI, H.; HEUNIS, A. J. (1997) 'Strong diffusion approximations for recursive stochastic algorithms', *IEEE Transactions on Information Theory* 43, 512–523.

[110] PUTERMAN, M. (1994) *Markov Decision Processes*, John Wiley, New York.

[111] ROBBINS, H.; MONRO, S. (1951) 'A stochastic approximation method', *Annals of Mathematical Statistics* 22, 400–407.

[112] ROCKAFELLAR, R. T. (1970) *Convex Analysis*, Princeton Univ. Press, Princeton, N.J.

[113] RUBINSTEIN, R. (1981) *Simulation and the Monte Carlo Method*, John Wiley, New York.

[114] RUDIN, W. (1986) *Real and Complex Analysis* (3rd ed.), McGraw-Hill, New York.

[115] RUDIN, W. (1991) *Functional Analysis* (2nd ed.), McGraw-Hill, New York.

[116] RUPPERT, D. (1988) 'A Newton-Raphson version of the multivariate Robbins-Monro procedure', *Annals of Statistics* 13, 236–245.

[117] RUSZCZYNSKI, A.; SYSKI, W. (1983) 'Stochastic approximation method with gradient averaging for unconstrained problems', *IEEE Transactions on Automatic Control* 28, 1097–1105.

[118] SANDHOLM, W. (1998) 'An evolutionary approach to congestion', Discussion paper available at: http://www.kellogg.northwestern.edu/research/math/papers/1198.pdf.

[119] SARGENT, T. (1993) *Bounded Rationality in Macroeconomics*, Clarendon Press, Oxford, UK.

[120] SASTRY, P. S.; MAGESH, M.; UNNIKRISHNAN, K. P. (2002) 'Two timescale analysis of the Alopex algorithm for optimization', *Neural Computation* 14, 2729–2750.

[121] SASTRY, P. S.; PHANSALKAR, V. V.; THATHACHAR, M. A. L. (1994) 'Decentralized learning of Nash equilibria in multi-person stochastic games with incomplete information', *IEEE Transactions on Systems, Man and Cybernetics* 24, 769–777.

[122] SHAMMA, J. S.; ARSLAN, G. (2005) 'Dynamic fictitious play, dynamic gradient play, and distributed convergence to Nash equilibria', *IEEE Transactions on Automatic Control* 50, 312–327.

[123] SINGH, S. P.; KEARNS, M.; MANSOUR, Y. (2000) 'Nash convergence of gradient dynamics in general-sum games', *Proc. of the 16th Conference on Uncertainty in Artificial Intelligence*, Stanford, Calif., 541–548.

[124] SMITH, H. (1995) *Monotone Dynamical Systems*, American Math. Society, Providence, R.I.

[125] SPALL, J. C. (1992) 'Multivariate stochastic approximation using a simultaneous perturbation gradient approximation', *IEEE Transactions on Automatic Control* 37, 332–341.

[126] SPALL, J. C. (2003) *Introduction to Stochastic Search and Optimization*, John Wiley, Hoboken, N.J.

[127] STROOCK, D. W.; VARADHAN, S. R. S. (1979) *Multidimensional Diffusion Processes*, Springer Verlag, New York.

[128] TSITSIKLIS, J. N. (1994) 'Asynchronous stochastic approximation and Q-learning', *Machine Learning* 16, 185–202.

[129] TSITSIKLIS, J. N.; VAN ROY, B. (1997) 'An analysis of temporal-difference learning with function approximation', *IEEE Transactions on Automatic Control* 42, 674–690.

[130] VEGA-REDONDO, F. (1995) *Evolution, Games and Economic Behaviour*, Oxford Univ. Press, Oxford, UK.

[131] WAGNER, D. H. (1977) 'Survey of measurable selection theorems', *SIAM Journal on Control and Optimization* 15, 859–903.

[132] WASAN, M. (1969) *Stochastic Approximation*, Cambridge Univ. Press, Cambridge, UK.

[133] WATKINS, C. I. C. H. (1988) '*Learning from Delayed Rewards*', Ph.D. thesis, Cambridge Univ., Cambridge, UK.

[134] WILSON, F. W. (1969) 'Smoothing derivatives of functions and applications', *Transactions of the American Math. Society* 139, 413–428.

[135] WILLIAMS, D. (1991) *Probability with Martingales*, Cambridge Univ. Press, Cambridge, UK.

[136] WONG, E. (1971) 'Representation of martingales, quadratic variation and applications', *SIAM Journal on Control and Optimization* 9, 621–633.

[137] YAN, W.-Y.; HELMKE, U.; MOORE, J. B. (1994) 'Global analysis of Oja's flow for neural networks', *IEEE Transactions on Neural Networks* 5, 674–683.

[138] YOSHIZAWA, S.; HELMKE, U.; STARKOV, K. (2001) 'Convergence analysis for principal component flows', *International Journal of Applied Mathematics and Computer Science* 11, 223–236.

[139] YOUNG, H. P (1998) *Individual Strategy and Social Structure*, Princeton Univ. Press, Princeton, N.J.

[140] YOUNG, H. P. (2004) *Strategic Learning and Its Limits*, Oxford Univ. Press, Oxford, UK.

Index

Bibliographical note

Chapter 1: The urn example is loosely adapted from Arthur (1994). The material on 'least mean square' algorithm is standard, see, e.g., Benveniste, Metivier and Priouret (1990), among others.

Chapter 2: The main result is adapted from Benaim (1999).

Chapter 3: This is based on Borkar and Meyn (2000) and Abounadi, Bertsekas and Borkar (2002).

Chapter 4: The treatment of lock-in probability is along the lines of Borkar (2002). The estimates based on Burkholder inequality, however, are older and can be found in Benveniste, Metivier and Priouret (1990) and Benaim (1999). The 'avoidance of traps' result follows Borkar (2003), but is only one among many such, see the end of this chapter for more references.

Chapter 5: This is adapted from Benaim, Hofbauer and Sorin (2005).

Chapter 6: The two timescale scheme follows Borkar (1996). The treatment of natural timescales follows Borkar (2006), which extends to 'controlled Markov' noise the classical results for 'Markov noise' taking a somewhat different approach. The latter can be found, e.g., in Benveniste, Metivier and Priouret (1990) and Kushner and Yin (2003).

Chapter 7: This follows Borkar (1998). For a somewhat different treatment, see Kushner and Yin (1987a, 1987b, 2003).

Chapter 8: This works out a rather simple special case of the much more general result found, e.g., in Benveniste, Metivier and Priouret (1990).

Chapter 9: There is scatterred work on constant stepsize algorithms. This chapter does not closely follow any particular one, except Borkar and Meyn (2000) in the early part.

Chapters 10 and 11: The (many) references for these chapters are given within the chapters themselves.

Printed in the United States
By Bookmasters